STARGAZING ACROSS CULTURES:

Exploring Indigenous Astronomy and Traditions

STARGAZING ACROSS CULTURES:

Exploring Indigenous Astronomy and Traditions

AUTHORS:

AUSTIN MARDON
ELINE EL-AWAD GONZALEZ
HAFSAH SAAJIDH
HASSAN OUDAH
JEREMY STEEN
ODETTE WILLS
YASMEEN EL-IRANI

EDITOR:

CATHERINE MARDON

GM★ PRESS

First Printing: 2023

Typeset and Cover Design by Josh Harnack

ISBN: 978-1-77889-051-2
eBook ISBN: 978-1-77889-052-9

Golden Meteorite Press
103 11919 82 St NW
Edmonton, AB T5B 2W3
www.goldenmeteoritepress.com

TABLE OF CONTENTS

CONSTELLATIONS AND STAR STORIES 9
LUNAR CYCLES AND CEREMONIES 19
SUN AND SEASONAL CYCLES 32
NAVIGATION AND ORIENTATION 41
INDIGENOUS ASTRONOMY AND SCIENCE 52
WESTERN KNOWLEDGE AND TRADITIONAL KNOWLEDGE .63

My eyes are fixed through the hole of a microscope. I have never been fascinated by astronomy before. Perhaps you are an asteroid. You are mighty and rough; you can be dangerous and cause damage. There is a lot more to see in the sky. Let each star in the sky be a reminder of an ancient and insightful thought. Is it not beautiful? You, the reader, can also be a star. A star is knowledgeable and insightful. As you read this book, have an open mind for traditional knowledge. The writers of this book not only want to shine a light on research but to shine a light on stories that have been told for a long time.

- Odette Wills

CHAPTER ONE

CONSTELLATIONS AND STAR STORIES

By Eline El-Awad Gonzalez

INTRODUCTION

Aboriginal astronomy is a fascinating and complex field of study that explores the astronomical knowledge and traditions of Indigenous cultures (Norris, 2016). For tens of thousands of years, the Indigenous have developed a deep understanding of the night sky and its significance in their lives, using it for navigation, seasonal tracking, and storytelling. This ancient knowledge is rooted in spiritual and cultural beliefs that have been passed down through generations of Indigenous communities. Despite the devastating impacts of colonization and cultural suppression, many Indigenous have maintained their connection to the stars, planets, and celestial bodies, preserving their traditional knowledge for future generations (Steffens, 2009). Today, the study of Aboriginal astronomy is helping to shed new light on the rich cultural heritage of Indigenous peoples, while also contributing to our broader understanding of astronomy and human history.

Aboriginal astronomy refers to the traditional astronomical knowledge and practices of the Indigenous. This includes their understanding of the Sun, Moon, stars, planets, and the Milky Way, as well as their movements in the sky (Nakata et al., 2014). The knowledge of astronomy has been passed down orally, through ceremonies, and various forms of artwork, and is an integral part of their cultural, mythological, and religious beliefs. Indigenous people have a deep understanding of the celestial bodies and their motions, which allowed them to develop practical uses for them such as creating calendars and navigating across the continent and waters (Australian Geographic, 2010). There are diverse astronomical traditions among different Indigenous communities, each with its unique cosmology, but common themes and systems can be observed between groups.

The constellations in Aboriginal astronomy are named based on their shapes, similar to traditional Western astronomy, such as the Pleiades, Orion, and the Milky Way (Norris, 2016). However, there are also unique constellations such as Emu in the Sky, which describes the dark patches instead of the points lit by the stars (D'Arcy, 1994). Contemporary Indigenous art often references astronomical subjects and related lore, including the Seven Sisters. Given the long history of Aboriginal astronomy, the Indigenous are often recognized as the world's first astronomers (Nakata et al., 2014). The study of Aboriginal astronomy provides a glimpse into the rich cultural heritage of Indigenous people and contributes to our broader understanding of astronomy and human history.

EMU IN THE SKY

The "Emu in the Sky" is a prominent constellation in Aboriginal astronomy that features dark nebulae visible against the Milky Way background (Steffens, 2009). The Emu's head is represented by the dark Coalsack nebula located near the Southern Cross, while its body and legs are an extension of the Great Rift trailing out to Scorpius (National Science and Technology Centre, n.d.). This constellation is widely used in Aboriginal culture and is depicted in rock engravings found in Ku-ring-gai Chase National Park, which include representations of the creator-hero Daramulan and his emu-wife.

The Coalsack nebula, which forms the head of the Emu in the Sky constellation, has a different cultural significance to the Wardaman people, who view it as the head of a lawman (P'Arcy, 1994). The Wiradjuri people refer to the constellation as Gugurmin, which is the inspiration for the title of Bruce Pascoe's book, Dark Emu. In May 2020, the Royal Mint launched a limited edition commemorative one-dollar coin as the first in its "Star Dreaming" series, celebrating Indigenous astrology (Aboriginal Astronomy, n.d.). The significance of the Emu in the Sky constellation and other celestial bodies in Aboriginal astronomy highlights the rich cultural heritage of the Indigenous and the importance of preserving this knowledge for future generations.

CANOE IN ORION

The Yolŋu people have a unique interpretation of the constellation Orion, which they call Julpan or Djulpan, representing a canoe (Wild Walks, n.d.). The Yolŋu tell the story of three brothers who went fishing and broke their law by eating a forbidden sawfish. In response, the Sun-woman, Walu, created a waterspout that lifted them and their canoe into the sky. The three stars at the centre of the constellation, which represent Orion's Belt in Western mythology, are the three brothers. The Orion Nebula, located above the Belt, represents the forbidden fish, and the bright stars Betelgeuse and Rigel represent the bow and stern of the canoe.

This Yolŋu interpretation of Orion highlights the integration of astronomy into their cultural beliefs and practices (P'Arcy, 1994). The astronomical story of the three brothers and their forbidden act underpins the ethical and social codes that guide Yolŋu's behaviour on Earth. This is just one example of how astronomical legends are interwoven with cultural beliefs and values and used to explain natural phenomena and societal norms. Such stories are essential to understanding the deep connection between Indigenous peoples and their environments and to appreciating the rich heritage of Indigenous astronomy.

PLEIADES

The Pleiades constellation figures in the Dreamings and songlines of several Aboriginal peoples, usually referred to as the seven sisters (Steffens, 2009). The story has been described as "one of the most defining and predominant meta-narratives chronicled in ancient mainland Australia", which describes a male ancestral being (with names including Wati Nyiru, Yurlu and others) who pursues seven sisters across the middle of the Australian continent from west to east, where the sisters turn into stars. Told by many peoples across the country, using varying names for the characters, it starts in Martu country in the Pilbara region of Western Australia (specifically, Roebourne), and travels across the lands of the Ngaanyatjarra (WA) to (Anangu Pitjantjatjara Yankunytjatjara, or APY lands, of South Australia, where the Pitjantjatjara and Yankunytjatjara peoples live. The story also includes Warlpiri lands and the Tanami Desert of the Northern Territory.
The Yamatji people of the Wajarri language group, of the Murchison region in Western Australia, call the sisters Nyarluwarri (Nakata et

al., 2014). When the constellation is close to the horizon as the sun is setting, the people know that it's the right time to harvest emu eggs, and they also use the brightness of the stars to predict seasonal rainfall. In the Kimberley region of Western Australia, the eagle-hawk chases the seven sisters up into the sky, where they become the star cluster and he becomes the Southern Cross (Steffens, 2009).

In the Western Desert cultural bloc in central Australia, they are said to be seven sisters fleeing from the unwelcome attentions of a man represented by some of the stars in Orion, the hunter (Aboriginal Astronomy, n.d.). In these stories, the man is called Nyiru or Nirunja, and the Seven Sisters' songline is known as Kungkarangkalpa. The seven sisters' story often features in the artwork of the region, such as the 2017 painting by Tjungkara Ken, Kaylene Whiskey's 2018 work "Seven Sistas", and the large-scale installation by the Tjanpi Desert Weavers commissioned as a feature of the National Gallery of Australia's 2020 Know My Name Exhibition. The Museum of Contemporary Art Australia in Sydney holds a 2013 work by the Tjanpi Desert Weavers called Minyma Punu Kungkarangkalpa (Seven Sisters Tree Women) (Nakata et al., 2014). In March 2013, senior desert dancers from the APY Lands (South Australia) in collaboration with the Australian National University's ARC Linkage and mounted by artistic director Wesley Enoch, performed Kungkarangkalpa: The Seven Sisters Songline on the shores of Lake Burley Griffin in Canberra.

In the Warlpiri version of the story, the Napaljarri sisters are often represented carrying a man called Wardilyka, who is in love with the woman (Steffens, 2009). But the morning star, Jukurra-jukurra, a man from a different skin group who is also in love with the sisters, chases them across the sky. Each night they launch themselves into the sky, and each night he follows them. This story is known as the Napaljarri-warn Jukurrpa. The people around Lake Eyre tell how the ancestor male is prevented from capturing one of the seven sisters by a great flood (Australian Geographic, 2010). The Wirangu people have a creation story embodied in a songline of great significance based on the Pleiades. In the story, the hunter (the Orion constellation) is named Tgilby. Tgilby, after falling in love with the seven sisters, known as Yugarilya, chases them out of the sky, onto and across the earth. He chases them as the Yugarilya chase a snake, Dyunu.

The Boonwurrung people of the Kulin nation of Victoria tell the Karatgurk story, which tells of how a crow robbed the seven sisters of their secret of how to make fire, thus bringing the skill to the people on earth (Steffens, 2009). In another story, told by people of New South Wales, the seven sisters are beautiful women known as the Maya-Mayi, two of whom are kidnapped by a warrior, Warrumma, or Warunna. They eventually escape by climbing a pine tree that continually grows up into the sky where they join their other sisters.

THE MILKY WAY

The Kaurna people who inhabited the Adelaide Plains had a name for the Milky Way: wodliparri, which translates to "house river" in the Kaurna language (Australian Geographic, 2010). They believed that the River Torrens, known as Karrawirra Parri, was a reflection of this mystical house river. The Yolŋu people held the belief that when they died, they were transported to the spirit-island Baralku in the sky by a mystical canoe called Larrpan (Norris, 2016). Their campfires could be seen burning along the edge of the Milky Way, which they believed was a great river. The canoe returned to Earth as a shooting star, indicating to their loved ones that they had arrived safely in the spirit-land. The Aboriginals also believed that the canoe was a representation of a god.

SUN AND MOON

According to the Yolŋu people, the Sun-woman, Walu, lights a fire each morning, which creates the dawn (National Science and Technology Centre, n.d.). She paints herself with red ochre, which spills onto the clouds, creating the sunrise. Carrying a torch, she moves across the sky from east to west, creating daylight. When she descends from the sky at the end of her journey, some of her ochres rub off onto the clouds, creating the sunset (Aboriginal Astronomy, n.d.). She then puts out her torch and travels back to her starting camp in the east underground throughout the night. The Yolŋu people also know Walu by the name Wuriupranili. Other stories about the Sun include Wala, Yhi, and Gnowee.

According to Yolŋu legend, the Moon-man, Ngalindi, was once young and slim (the waxing Moon) but became fat and lazy (the full Moon) (National Science and Technology Centre, n.d.). To escape his wives who wanted to chop bits off him with their axes (the waning Moon), he climbed a tall tree towards the Sun but died from his wounds (the new Moon). After three days, he rose again to repeat the cycle, which continues to this day. The Kuwema people from the Northern Territory claim that he grows fat at every full Moon by devouring the spirits of those who disobey tribal laws. Another story by the Aboriginals of Cape York involves the creation of a giant boomerang thrown into the sky, which becomes the Moon. In Southern Victoria, there is a legend about a beautiful woman who is forced to live in the sky by herself following a series of scandalous affairs. The Yolŋu people also associate the Moon with the tides.

ECLIPSES

According to the Warlpiri people, a solar eclipse occurs when the Moon-man hides the Sun-woman while making love to her (Nakata et al., 2014). This explanation is shared by other groups, including the Wirangu. The Ku-ring-gai Chase National Park features several engravings of a crescent shape with sharp horns pointing down and a drawing of a man and woman below it. Although most researchers assume that the crescent shape represents a boomerang, some argue that it is more easily interpreted as a depiction of a solar eclipse, with the man-and-woman explanation illustrated below it.

VENUS

The Yolŋu people observe an important ceremony to mark the rising of Venus, which they refer to as Barnumbirr, meaning "Morning Star and Evening Star" (Norris, 2016). After sunset, they gather to witness the planet's appearance, and as it rises in the early hours before dawn, they believe that it pulls a rope of light behind it that is connected to the island of Baralku on Earth. Using a Morning Star Pole that is richly decorated, they can communicate with their deceased loved ones along this rope, demonstrating that they still love and remember them. This belief system reflects the strong connection that the Indigenous have with their ancestors, and the belief in an afterlife that is accessible through the natural world (Kendall, 2020). The Morning Star Pole is a powerful symbol of this connection, serving as a conduit between the living and the dead. This ceremony shows how the Indigenous have a deep understanding of the natural world and use it to express their spirituality and connect with their past.

ETA CARINAE

In 2010, Macquarie University astronomers Duane Hamacher and David Frew discovered that the Boorong Aboriginal people witnessed the outburst of Eta Carinae in the 1840s and incorporated it into their oral traditions (Wild Walks, n.d.). They referred to it as Collowgulloric War, the wife of War (Canopus, the Crow - waː). This is the only indigenous record of Eta Carinae's outburst that has been definitively identified in the literature so far.

ASTRONOMICAL CALENDARS

Aboriginal calendars differ from the European calendar, with many groups using a six-season calendar and marking the seasons by the stars visible during them. The Pitjantjatjara, for instance, recognize the start of winter when the Pleiades rise at dawn in May (Aboriginal Astronomy, n.d.). The rising or setting of stars or constellations indicates to Aboriginals when it is time to move to a new place or seek out a new food source. For instance, the Boorong people in Victoria gather eggs when the Malleefowl disappears in October to "sit with the Sun," and other groups know when the dingo puppies are about to be born when Orion appears in the sky.4 The Yolŋu know that the Macassan fishermen will soon arrive to fish for trepang when Scorpius appears.

It is not clear to what extent Aboriginal people were interested in the precise motion of celestial bodies, although some of the stone arrangements in Victoria, such as Wurdi Youang, may have been used to predict and confirm the equinoxes and solstices (P'Arcy, 1994). This arrangement aligns with the setting sun at the solstices and equinox, but its age is unknown. The Nganguraku people have rock engravings at Ngaut Ngaut that represent lunar cycles, but much of their culture, including their language, was lost due to the ban on such things by Christian missionaries before 1913.

CONTEMPORARY CULTURE

Contemporary Aboriginal art frequently showcases an astronomical theme that reflects the astronomical elements of the artists' cultures (Pascoe, 2014). Artists such as Gulumbu Yunupingu, Bill Yidumduma Harney, and Nami Maymuru have received recognition for their work, including awards and being finalists in the Telstra Indigenous Art

Awards. In 2009, a display of Indigenous Astronomical Art from WA, called Ilgarijiri, was introduced at AIATSIS in Canberra in conjunction with a Symposium on Aboriginal Astronomy.

The seven sisters are a popular theme in contemporary art. Gabriella Possum and Michelle Possum, daughters of the late Clifford Possum Tjapaltjarri, frequently paint the Seven Sisters Dreaming in their paintings Pascoe, 2014). They inherited this Dreaming through their maternal line, and their work highlights the importance of this tradition in their culture. Through their art, they celebrate their heritage and help to ensure that it is passed down to future generations.

IMPACT ON THE MODERN DAY

Indigenous astronomy has had a profound impact on our understanding of the cosmos and our place within it. For thousands of years, Indigenous peoples across the globe have developed complex systems for observing and interpreting celestial phenomena, often informed by their cultural, spiritual, and practical needs (Nakata et al, 2014). Despite centuries of colonization and marginalization, many Indigenous communities have continued to maintain and adapt these traditions, drawing on both ancient knowledge and modern science to expand our understanding of the universe.

One of the most significant contributions of Indigenous astronomy has been the recognition of the interconnectedness of all things. Many Indigenous cosmologies view the stars, planets, and other celestial bodies as living beings with their spirits and personalities, connected to the earth and its inhabitants in complex ways (Kendall, 2020). This holistic worldview challenges the Western scientific paradigm, which often separates humans from the natural world and emphasizes objectivity and rationality over subjectivity and emotion.

Furthermore, Indigenous astronomy has helped to challenge and expand our understanding of time and space (Steffens, 2009). Many Indigenous cultures have developed complex calendars based on the movements of the stars and planets, which not only mark the passage of time but also reflect cultural and ecological cycles. By recognizing the cyclical nature of time, Indigenous peoples have been able to develop sustainable ways of living that honour and protect the earth and its resources.

In today's world, Indigenous astronomy is increasingly recognized as a valuable source of knowledge and inspiration (Norris, 2016). Many Indigenous astronomers, educators, and activists are working to preserve and promote traditional astronomical practices, while also using them to engage with contemporary scientific research. Indigenous perspectives are being integrated into astronomy education, helping to broaden and deepen our understanding of the universe and its place in our lives.

However, it is important to note that the recognition and integration of Indigenous astronomy into mainstream scientific discourse is not without its challenges (Kendall, 2020). Many Indigenous communities continue to face ongoing colonization, marginalization, and erasure, which can make it difficult to preserve and share their knowledge. Additionally, Western scientific institutions and practices may not always be receptive to non-Western ways of knowing, which can perpetuate power imbalances and limit the potential for collaboration and mutual learning.

Nonetheless, Indigenous astronomy offers a valuable perspective that can help us to navigate the complex challenges facing our planet today (Pascoe, 2014). By recognizing the interconnectedness of all things, we can work to build more sustainable and equitable societies that honour and protect the earth and its diverse communities. By engaging with Indigenous knowledge and practices, we can expand our understanding of the cosmos and its place in our lives, while also fostering greater respect and understanding across cultural divides.

Overall, Indigenous astronomy has much to offer us in today's world, as we grapple with pressing environmental, social, and scientific issues. By recognizing and valuing Indigenous knowledge, we can broaden our understanding of the universe and our place within it, while also honouring the diverse cultures and perspectives that make up our global community.

REFERENCES

Norris, R. P. (2016). Dawes Review 5: Aboriginal Astronomy and Navigation. Publications of the Astronomical Society of Australia, 33, e039. https://doi.org/10.1017/pasa.2016.25

Steffens, M. (2009). Australia's first astronomers. ABC News. Retrieved from https://www.abc.net.au/news/2009-07-27/australias-first-astronomers/1362886

Nakata, M., Hamacher, D., Warren, J., Byrne, A., Pagnucco, M., Harley, R., Venugopal, S., Thorpe, K., Neville, R., Bolt, R. (2014) Using Modern Technologies to Capture and Share Indigenous Astronomical Knowledge. Australian Academic & Research Libraries, 45(2), 101-110. https://doi.org/10.1080/00048623.2014.917786

Australian Geographic. (2010). Aboriginal astronomers: World's oldest? Australian Geographic. Retrieved from https://www.australiangeographic.com.au/topics/history-culture/2010/05/aboriginal-astronomers-worlds-oldest/

D'Arcy, P. (1994). The Emu in the Sky: Stories about the Aboriginals and the day and night skies. In M. Sutton (Ed.), The National Science and Technology Centre (pp. 15-16). ISBN 978-0-64618-202-5.

The National Science and Technology Centre. (n.d.). Aboriginal Astronomy. Retrieved from https://www.questacon.edu.au/outreach/programs/aboriginal-astronomy/

Aboriginal Astronomy. (n.d.). Emu in the Sky. Retrieved from https://aboriginalastronomy.com.au/the-emu-in-the-sky/

Wild Walks. (n.d.). Elvina Bay Aboriginal Engraving Walk. Retrieved from https://www.wildwalks.com/bushwalking-and-hiking-in-nsw/kuring-gai-chase-national-park/elvina-bay-aboriginal-engraving-walk.html

Kendall, R. (2020). 65,000 years of star gazing for Indigenous Australians. Echonetdaily. Retrieved from https://www.echo.net.au/2020/10/65000-years-of-star-gazing-for-indigenous-australians/

Pascoe, B. (2014). Dark Emu: Black Seeds: Agriculture or Accident? Magabala Books. ISBN 978-1-922142-43-6.

CHAPTER TWO

LUNAR CYCLES AND CEREMONIES

By Yasmeen El-Irani

INTRODUCTION

Long before the Babylonians, Greeks, the Renaissance, and the Enlightenment existed the first astronomy: one rooted in and based on Indigenous knowledge. As it holds a more holistic approach and draws on truths and values from outside the sciences, Indigenous astronomy plays an important part in the history and culture [many] societies that date back centuries. It involves the observation and interpretation of the natural world, including the movements of celestial bodies [such as the moon]. In many Aboriginal traditions, the planets are seen as children of the Sun and Moon; and so, both the sun and the moon hold prominent places in the lives, beliefs, ceremonies, and understandings of the people.

In many indigenous cultures, lunar cycles and ceremonies play a crucial role in spiritual and cultural life. In this essay, we will explore traditional lunar calendars and their importance, the significance of lunar phases in ceremonies, and the connection between the moon and tides, plant growth, and hunting seasons. We will also delve deeper into the role of the Lunar calendar in holistic health and wellness, women's health and fertility, spiritual movements, and manifestation rituals.

While these beliefs and practices have been present for centuries, and based on oneness with the planets - knowledge that has emerged has been adapted and implemented in modern-day, Western notions of thinking, being, and living. And so, [lastly], we will be discussing the way Indigenous astronomy has been adapted in a global setting, as well as the concerns regarding its appropriation and commercialization.

TRADITIONAL LUNAR CALENDARS

Separating ourselves, as humans, from nature results in the tendency to see ourselves as superior, as well as believing that other living entities exist solely for our use and benefit. It is a general, global - but mostly Western - anthropological belief system that is deeply rooted in the capitalist obsession with consumerism. Rather than seeing land as a means to material, Indigenous tribes perceive land as The Great Spirit; they believe that there is a certain balance between the human and non-human world, making us one with the Earth.

However, the connection lies far beyond the soil we reside on, and the living beings that come forth from it. For thousands of years, humans have looked to the skied to understand our place in the universe. "Sky stories from the Siksika of the Blackfoot and the Ininewuk of the Cree reflect a distinct philosophy that helps guide relationships between individuals and the natural world. Traditional Ways of Knowing and astronomy knowledge are blended into stories that bind water, land, humans and animals into the regular rhythms of celestial movement." (Rothney, 2023)

Many indigenous cultures use lunar calendars to mark the passage of time. Unlike the solar calendar used in Western societies, which is based on the movement of the Earth around the sun, lunar calendars are based on the phases of the moon. The moon's cycles take an estimated 29.5 days to orbit the Earth. This means that each lunar cycle lasts about a month (each beginning and ending with the new moon), which is why many lunar calendars have 12 months.

The importance of traditional lunar calendars lies in their connection to the natural world, as they are often closely tied to seasonal changes, as well as the cycles of nature. Before scientists created tools like weather satellites, computer models, and sensors to collect and analyze data - as well as make future predictions - Indigenous cultures used the cycles of the moon to track changing seasons, predict weather patterns, and understand the behaviour of plants and animals. This knowledge is crucial for survival in many indigenous societies, where hunting, fishing, and agriculture are still important ways of life. It allows them to determine the right time to adjust their planting, harvesting, and other activities to be in balance, and one, with nature.

Another aspect of significance here is that: Indigenous cultures approach this method of understanding the natural world based on traditional ecological knowledge (TEK). TEK is a holistic and place-based knowledge system that is passed down through generations and rooted in direct observation and experience. Not only is it considered to be valuable, but also deeply sacred; it is a way of preserving and transmitting cultural traditions and practices that are rooted in spiritual beliefs. This knowledge is seen as a gift from the Creator or the natural world, and therefore something that should be treated with respect and reverence. Additionally, TEK is closely tied to indigenous peoples' sense of place and belonging. For many indigenous peoples, their knowledge of the land and the natural world is inseparable from their identity and their sense of cultural continuity.

While the lunar calendar is a spiritual and cultural symbol, the moon is often seen as a powerful and sacred entity, with each phase representing a different aspect of its power and energy. We will explore this in further sections in this chapter.

Despite global, human, and evolutional differences - every culture on the face of the Earth uniquely sees the moon, along with its own stories of the faces it shows us throughout the month. Although the lunar calendar plays a vital role in indigenous cultures by providing a framework for rhythm and flow of life - it is not limited to their practices. For 3,000 years, the lunar calendar served the purpose of guiding agricultural affairs; and so, many ancient civilizations have their lunar calendar. For example, the Chinese Lunar Calendar (which is most widely referred to in Asia) is based on a very intricate and complex calculation. "Contrary to its name, the Chinese lunar calendar isn't just based on the phases of the moon, but also the position of the sun in the sky." Thus, the year is divided into 24 solar terms, splitting it into 4 seasons of exactly 3 months each.

Whether it is TEK, Chinese calculations, Western science - or any other cultural method of obtaining and using knowledge - all provide a valuable way of understanding the natural world. The beauty of their differences arises from each of their representations of unique worldviews and approaches to knowledge, not to mention - a way of life.

LUNAR CYCLES & CEREMONIES

Aside from its practical uses, the lunar calendar also serves as an important spiritual and cultural symbol for various Indigenous cultures around the world; and so, throughout history, the moon's intrinsic power and energy have often been considered to be sacred. The moon has captivated the human imagination since time immemorial; it has been regarded with great reverence, respect, and sometimes even religious devotion. In many cultures and belief systems, the moon is considered to be a divine or mystical force, with its cycles and phases believed to hold profound spiritual significance. The moon's influence on the tides, the growth of crops, and even human emotions and behaviour (which we will discuss later) have also contributed to its perceived sacredness. As such, the moon's influence continues to captivate and inspire people worldwide, serving as a source of fascination and wonder.

Each phase of the moon represents a different aspect of this power and energy, making it a powerful force that continues to inspire awe in many people. There are generally eight phases of the moon; however, the significance of each varies among different indigenous cultures. Although differences are present, there are common interpretations of each lunar cycle. As well, ceremonies are held on specific dates that correspond to the lunar calendar (i.e. the season or natural phenomena that are associated with that time of year).

THE NEW MOON

The new moon occurs when the moon is not visible in the sky. As it is the first phase it marks the start of a new lunar cycle, in many Indigenous cultures, it is typically viewed as a symbolic representation of new beginnings, fresh starts, and setting intentions for the future. The power of thought is a significant concept in many indigenous cultures and is viewed as a tool for personal growth, healing, and spiritual development. Indigenous people believe that thoughts are a form of energy and that our thoughts can interact with the energy of the world around us. Thought is the initial seed that determines what results you will experience. For example: if you sow a seed into the ground, it will begin to bear fruit and multiply - it creates new crops of the seed initially sown. If you want a new fruit, you need to sow a new seed. In that same sense, you cannot think of poverty and experience abundance and prosperity - just as you cannot think of abundance and prosperity, then experience poverty. Thus, the power of thought is expressed in rituals, practices, and ceremonies such as meditation, prayer, and visualization; it serves as a way to manifest new energies and fresh starts during this time. Hence, this concept plays a very significant role during the new moon, as it brings about positive changes in their lives.

As this phase of the moon is viewed as a time of rebirth and growth, it is often associated with the element of water and emotions; and so, the new moon is not just a welcoming for positive energy, but also a purification and cleansing to release negative energy. Additionally, due to this association, the new moon is also seen as a time for the literal planting and harvesting of new seeds. From a practical perspective, planting during the new moon has also been linked to the gravitational pull of the moon. As the moon's gravity affects the tides, it is believed to also influence the moisture content of the soil, which can impact plant growth. Planting during the new moon is said to coincide with a time when the soil is most receptive to new seeds and has the best conditions for germination.

In terms of harvesting, the new moon is often associated with the end of one lunar cycle and the beginning of another. This is a time to reflect on the growth and changes that have occurred and to gather the fruits of one's labour. Indigenous cultures may hold ceremonies or rituals to express gratitude for the harvest and to honour the cycles of nature. These ceremonies can vary depending on the specific culture and region, but some common examples include harvest festivals, thanksgiving ceremonies, blessing ceremonies, and healing ceremonies.

THE WAXING CRESCENT MOON
The next moon phase is the waxing crescent moon; it occurs when a sliver of the moon becomes visible in the sky. With the growth of its visibility - it represents learning and gaining strength, as well as taking action and making progress towards your goals. In some indigenous cultures, this phase is also associated with the element of air and the mind, and it may be a time for mental clarity and creativity. Additionally, this phase is significant in the planting and harvesting of seeds that occurred during the new moon. The dark sky during the new moon represents the fertile ground in which new seeds can be planted and nurtured; however, the first sliver of the waxing crescent moon that

follows signifies the growth and potential of those seeds. In some cultures, the waxing crescent moon is celebrated with a "moon dance," where people gather to dance, sing, and make offerings to the spirits. This may be a time to honour ancestors or connect with spirit guides for guidance and support in manifesting one's goals.

FIRST QUARTER MOON

This is when half of the moon is visible in the sky, and it represents a time for decision-making and taking action. This is a time for making progress towards your goals and for overcoming obstacles and challenges, as the moon moves from new to full. As the sun and moon are in equal parts of the sky, indigenous cultures may view this phase as a time of balance; as well, it may be a time for finding harmony and equilibrium in your life. The equal parts of light and darkness on the moon during this phase are seen as a reflection of the need for balance between taking action and reflection.

WAXING GIBBOUS MOON

This is when most of the moon is visible in the sky; as it continues to grow towards fullness, it represents a time of abundance, preparation, and gratitude towards your blessings. This phase represents a time of increased energy, enthusiasm, and creativity, as the moon builds towards its full illumination. Preparation for the full moon is significant, as it has a heightened energy that it brings with it. Additionally, indigenous cultures may associate this phase with the element of fire and the spirit, and it may be a time for connecting with your inner fire and passion.

FULL MOON

It is a time of ultimate light and guidance in darkness, a time where the howls of wolves are heard for miles: the full moon. It happens when the moon is visible in the sky - in its entirety; it represents a time of completion, illumination, and - as aforementioned - heightened energy. Hence, indigenous cultures often view the full moon as a sacred time of great spiritual and cosmic power. Thus, it is also seen as the time and indicator for ritual, ceremony, and spiritual connection.

STRENGTH AND TRANSFORMATION

Since its energy is believed to be particularly potent and transformative, indigenous cultures often celebrate this time with a variety of ceremonies to harness the moon's energy - in turn, manifesting all their dreams and desires. For example, Native Americans perform a "Ghost Dance" during the full moon to connect with the spirits and honour the cycles of nature; it is performed to bring about a spiritual renewal of the world, in which the earth would be restored to its harmonious, natural state. The Ghost Dance movement was also a powerful expression of the desire to maintain cultural traditions in the face of colonial oppression; it was their hope resilience and resistance against the effects of colonization, displacement, and disease on Native American communities.

HARVESTING, GRATITUDE, AND ABUNDANCE

Other cultures may hold ceremonies focused on harvest, as well as offering gratitude for the abundance in their lives. For example, the Anishinaabe people, an indigenous group native to the Great Lakes region of North America, have a full moon ceremony known as the "Manoominike-Giizis" or "Wild Rice Moon". This ceremony is held during the full moon of August or September, coinciding with the time of the wild rice harvest. During the ceremony, the Anishinaabe people offer thanks to the spirits of the wild rice, the water, and the land for providing them with sustenance. The ceremony begins with a feast where traditional foods, including wild rice, are served. Participants then gather around a sacred fire to offer prayers and sing traditional songs.

After the prayers and songs, participants enter the water to collect wild rice. They use traditional harvesting methods, which involve gently knocking the rice grains into a canoe with two wooden sticks. The rice is then prepared for storage or immediate consumption. The Manoominike-Giizis ceremony is an important part of Anishinaabe culture, as wild rice is considered a sacred food that is deeply connected to the land and the spirits. It is believed that this ceremony ensures a successful harvest, and honours the Anishinaabe people's relationship with the natural world.

This time is also considered sacred because this energy is seen as a source of abundance and prosperity, both materially and spiritually. the full moon is often associated with abundance because it represents the culmination of the lunar cycle when the energy of growth and expansion reaches its peak. Hence full moon ceremonies are held to honour this energy, as it marks the completion of a lunar cycle - and the beginning of a new one.

ROLE IN WOMEN'S HEALTH AND FERTILITY

The moon has been associated with female fertility and health for centuries. One reason why the full moon is associated with fertility is that it marks the midpoint of the lunar cycle, which is seen as a time of growth and expansion. Furthermore, in many cultures, the full moon is associated with the feminine principle and is viewed as a symbol of the creative power of the universe. It is often linked with the goddess or divine feminine energy, which is believed to be the source of all life and creation.

Additionally, the lunar cycle is often believed to be closely connected to the menstrual cycle; and so, women's bodies are seen as being particularly sensitive to the moon's phases. The full moon is often associated with ovulation and fertility, and some women may find that their menstrual cycle follows the lunar cycle, with menstruation occurring around the new moon and ovulation occurring around the full moon.

While the relationship between the moon and female health is not fully understood, many women continue to find solace and connection in the rhythms of the lunar cycle, and the moon remains an important symbol of feminine power and intuition. Interestingly enough, it has also been linked to their emotional and mental health. Some studies have suggested that women may be more prone to mood disturbances and sleep disruptions during certain phases of the moon, particularly the full moon. Evidence also suggests that the lunar cycle may have a profound impact on women's overall health and well-being, as it affects the levels of hormones such as melatonin.

WANING GIBBOUS MOON

This is when most of the moon is visible in the sky, but it is starting to shrink. It represents the release of negative energy, and letting go of what no longer serves us. This is a time for expressing gratitude for what you have accomplished - to make space for new growth and opportunities.

Indigenous cultures may view this phase as a time of transformation, as the old is released and the new is welcomed in.

THIRD QUARTER MOON
This is when half of the moon is visible in the sky, but it is starting to shrink. It represents reflection, evaluation, and re-adjustment. This is a time for taking stock of your progress and re-aligning with what resonates with you on a spiritual level. Indigenous cultures may see this as a phase for healing and renewal, as you evaluate what is working in your life and what needs to be released or changed.

WANING CRESCENT MOON
Lastly: the waning crescent moon; occurs when a sliver of the moon is visible in the sky, but is starting to shrink. It represents rest, rejuvenation, and preparation for the next lunar cycle: the new moon. This time is an indicator to slow down, take care of yourself, and prepare for new beginnings. Indigenous cultures may associate this phase with the element of earth and the physical body, hence why the elements of self-care, relaxation, and grounding are emphasized.

TIDES, PLANT GROWTH, AND HUNTING SEASON
In many indigenous cultures, the moon's gravitational pull is closely tied to natural phenomena such as tides, plant growth, and hunting seasons. The moon's gravity affects the Earth's oceans, causing the tides to rise and fall. This is because the moon's gravitational pull is strongest on the side of the Earth that is facing the moon, and weakest on the opposite side. As the Earth rotates, this creates a tidal bulge on both sides of the planet, which causes the ocean tides to rise and fall.

It is often considered that the connection between the moon and tides is reflected in the use of lunar calendars for fishing and hunting. For example, the Maori people of New Zealand use a lunar calendar to track the movements of fish and shellfish. They believe that certain phases of the moon are better for fishing, as the tides are more favourable during these times. The moon's gravitational pull also affects plant growth. In particular, the gravitational pull of the moon affects the movement of water in plants, which is known as transpiration. During the full moon, the gravitational pull of the moon on the Earth's water is stronger, which can cause water to rise in the soil and be absorbed by plants more easily. This can lead to increased growth and productivity in certain crops - hence, the celebration and ceremonies discussed in previous sections.

Additionally, the connection between the moon and plant growth is reflected in the use of lunar calendars for agriculture. For example, as aforementioned, the Anishinaabe people of North America use a lunar calendar to determine the best time for planting and harvesting rice. They believe that certain phases of the moon are better for planting, as the gravitational pull of the moon helps to stimulate growth.

The connection between the moon and hunting seasons is also reflected in indigenous culture and belief systems. In particular, the cycles of the moon are believed to have an impact on the behaviour of animals, such as deer and elk, that are hunted for food. For example, the Inuit people of Canada and Greenland use a lunar calendar to track the movements of caribou, which are a crucial source of food and clothing. They believe that during the new moon and full moon phases, the caribou are more active and easier to track, making these times ideal for hunting.

INTERSECT, OVERLAP, AND ADAPTATION TO WESTERN CULTURE

Many of the beliefs and practices associated with lunar cycles and ceremonies in indigenous cultures have been adapted and adopted by modern-day Western cultures, particularly in the areas of health and wellness, spirituality, and environmentalism. This has had a significant impact on modern-day Western notions of thinking, being, and living. They have inspired new ways of understanding and relating to the natural world, and have helped to promote more sustainable, holistic, and spiritually-connected ways of living.

In addition, the study of indigenous astronomy and traditions has helped to challenge Western notions of scientific knowledge and methodology. It has highlighted the importance of traditional knowledge and practices and has shown that there are multiple ways of understanding and interpreting the natural world. For example, the idea that the phases of the moon can affect human behaviour and emotions has gained popularity in recent years, particularly in the field of holistic health and wellness. Many people now use lunar calendars to track their menstrual cycles, as well as to plan and time various activities, such as meditation, yoga, and self-care practices.

Similarly, the connection between the moon and spiritual practices has been recognized by many Western religions and spiritual movements, such as Wicca and New Age spirituality. As mentioned, the full moon, in particular, is often associated with spiritual transformation, abundance, and manifestation; and so, is celebrated with rituals and ceremonies.

The indigenous understanding of the connection between the moon and the natural world has also influenced Western notions of environmentalism and conservation. Many people now recognize the importance of understanding and respecting the cycles and rhythms of the natural world, and of adopting sustainable practices that promote the health and well-being of the planet.

NEW WAVE OF SPIRITUALITY

The COVID-19 pandemic has had a profound impact on people's lives, leading many to question their beliefs and search for meaning and purpose in life. This has contributed to a surge in interest in spirituality and a renewed focus on personal growth and well-being. Individuals chose to share and decimate information on Tiktok - a social media platform that became popular among Gen Z and millennials during the pandemic. Many TikTok users, particularly those interested in spirituality, wellness, and alternative lifestyles, have used the platform to share their experiences and knowledge of indigenous practices related to the moon.

One way that these beliefs and practices have been adapted and marketed on TikTok is through the use of "moon rituals" and other lunar-themed content – hashtag-ing it #MoonTok. Many TikTok users have posted videos of themselves performing rituals and ceremonies associated with the phases of the moon, such as setting intentions during the new moon or releasing negative energy during the full moon. Additionally, they have also shared tips for incorporating lunar cycles into daily life, such as using lunar calendars to track menstrual cycles or setting intentions based on the phase of the moon.

The wave of people involved shared information about the meanings and symbolism associated with different lunar phases and cycles, such as the significance of the "blood moon" or the "blue moon." These beliefs and practices have been marketed and normalized on TikTok through the use of hashtags and other social media marketing strategies. Many TikTok users have created branded content around lunar cycles and ceremonies, promoting products and services related to wellness, spirituality, and alternative lifestyles. This, in turn, benefits them as individuals (in a material sense), rather than benefiting the whole (in a spiritual sense).

Thus, it is important to note that the sharing and dissemination of indigenous practices on TikTok are not without controversy. Some indigenous people have criticized the appropriation and commodification of their cultural practices by non-indigenous people, arguing that these practices should be respected and protected as part of their cultural heritage.

REFERENCES

Luna. Omnithreads. life. (n.d.). Retrieved from https://www.omnithreads.life/pages/luna

Mathematics, Moon Phases, and Tides. Indigenous Knowledge Institute. (n.d.). Retrieved from https://indigenousknowledge.unimelb.edu.au/curriculum/resources/mathematics,-moon-phases,-and-tides

MUSKRAT Magazine. (2016, June 28). Indigenous calendars mark much more than the spring equinox. MUSKRAT Magazine. Retrieved from http://muskratmagazine.com/indigenous-calendars-mark-much-more-than-the-spring-equinox/

Noon, K., & Hamacher, D. (2022, September 7). From the vault: Lunar traditions of the first Australians. Cosmos. Retrieved from https://cosmosmagazine.com/space/lunar-traditions-of-the-first-australians-2/

Rao Indigenous Skies. Faculty of Science. (2023, February 16). Retrieved from https://science.ucalgary.ca/rothney-observatory/community/Indigenous%20Skies

Rhonda. (2022, March 15). The lunar calendar explained. Pimachiowin Aki. Retrieved from https://pimaki.ca/the-lunar-calendar-explained/

Teacherkyla. (2019, June 6). Importance of the Moon in First Nations cultures across Canada. Teacher Kyla's Education Blog. Retrieved from https://teacherkyla.wordpress.com/2019/01/29/moon-phases-from-different-first-nations-perspectives/

Xie, A. (2018, October 3). Why does the lunar calendar still matter in Asian culture? Medium. Retrieved from https://angelxie.medium.com/why-does-lunar-calendar-still-matter-in-asian-culture-f5e054e107e6#:~:text=For%203%2C000%20years%2C%20the%20lunar,seasons%20of%20exactly%20three%20months.

CHAPTER THREE

SUN AND SEASONAL CYCLES

By Hassan Oudah

INTRODUCTION

Time, as one knows, revolves around the sun. Waking up in the morning relates to the sun that has risen, the day passing by, and it is over when the sun has set, and the night begins. The presence of the sun has an importance in nearly every culture that dates back to the very beginning of time. It is estimated that the sun began its formation through a molecular cloud mainly consisting of helium and hydrogen nearly 4.5 billion years ago. It was believed that the sun revolved around the earth, but a discovery made by mathematician Nicolaus Copernicus determined that the sun is at the centre of the universe and all other planets orbited around the sun in a circular direction. Earth is very dependent on the sun as it guides seasons, and climates and produces the possibility of plant life through photosynthesis. Seasonal cycles as they are known traditionally are four different seasons throughout the year. They are mostly known to be differentiated by the particular climate condition occurring at the time of the year. The sun and the seasonal cycle influences the way some cultures guide their lifestyle and traditions. For Indigenous communities, the sun has an essential role in traditions and ceremonies. Seasonal cycles are known to play a crucial role in the way Indigenous tribes live their lives as they historically had a great impact on their traditions and how they navigate life based on the changes occurring throughout the year. This chapter will discuss how the traditional solar calendar came about, the role of the sun in different Indigenous tribes' ceremonies and traditions and how some of the ceremonies are performed. As well as the significance of seasonal changes for Indigenous communities.

SOLAR CALENDAR

A major influence of the sun is the creation of the solar calendar. The sun began to have its influence over time as it began marking time by seasons. The solar calendar is how long it takes the Earth to revolve around the sun and finish the seasons. For the earth to move around the sun and return to the spot it started at; as perceived by the earth, is accounted for 365 days. Those 365 days consist of four different seasons within the seasonal cycle. The solar calendar aimed to follow the tropical year so that the seasons are occurring at the same time and are known well yearly. Around 2050 BCE, the Egyptians achieved this calendar by calculating it to be 365.25 days. However, it was not possible to account for 365.25 days as a quarter of a day is not possible. Nowadays, two main types of calendars are widely used, the solar calendar and the lunar calendar. Most cultures around the world use the solar calendar as the four seasons are correlated to the yearly rotation around the sun. The use of the lunar calendar is still present in traditional Chinese and Jewish calendars and practices.

INFLUENCE OF THE SUN

The sun's influence is present in nearly every culture and its importance can be observed in ancient civilizations as well. Around the time of 1000 to 1200 A.D., the ancient civilization of the Mayans in Mexico built the Pyramid of Kukulkan in the relation to the movement of the sun. The northwest and southwest corners have axes passing by them as they are positioned towards the rising point of the sun during the summer solstice and its setting point at the winter solstice. As for the Indigenous cultures, the sun has a crucial role in their spirituality and daily lives. "The sun is the most powerful physical presence in our lives." (Kidwell, 2020) Many actions and decisions were likely determined by the sun as it is perceived to be the guidance through life. The presence of the sun indicates the light that is meant to steer a person to a happy destination. The symbolism of the sun in life is said to be rooted in spirituality. Simply spotting the light from the sun does not automatically entitle one to a happy destination. The importance is that the symbolism of the sun encourages one to stick to a spiritual path that will bring happiness. The importance of the sun amongst Indigenous people is so prevalent that almost every tribe has its symbol of the sun that they portray through geometrical shapes and forms.

CEREMONIES AND TRADITIONS RELATING TO THE SUN

Culturally, the sun plays an essential role in ceremonies and traditions for a variety of tribes and nations. Across North America, many of the Indigenous tribes partake in the sacred 'Sun Dance' ceremonies. Different tribes practice different versions of the Sun Dance. The origin of the Sun Dance name has been obtained from the Oceti Sakowin tribe as 'Wi Wanyang Wacipi' which is translated to the 'Sun Gazing Dance.' Around the grasslands of Saskatchewan, Canada, the Indigenous nations have their way of celebrating the sacred ceremony. Since the sun is seen as an ultimate power source, the Sun Dance includes prayers where the participants offer prayers for the renewal of health and growth of the tribes' resources. Sun Dance ceremonies often last from four to eight days and they usually occur at any time between the early days of spring to the middle of the summer season. The main activity of the Sun Dance will be to observe the security within the community, the resumption of life, the great quality of the seasons and sound health. The ceremony will begin with the host of the ceremony selecting a centre pole. The centre pole will be brought in by a group of males who have been selected for the task. The Sun Dance will take place in the 'Sun Dance Lodge,' the lodge will be built circularly and the entry to it will be facing the east side to designate the entry of light. The pole that has been brought in must not touch the ground until it has been set down in the centre of the lodge. The centre pole will have a Thunderbird nest resting on top. This nest is a remarkable symbol as it represents the Eagle. The Eagle is known to be the messenger for the prayers sent to the Great Spirit. During the ceremony, the hosts and dancers would be dancing in shifts for the upcoming days. The dancers always face the centre pole toward the honoured spot of the host. They step to the drumbeats and say their prayers. This ceremony is the tribe's way to acknowledge and instil their belief in their sacred ways.

Oftentimes, the Sun Dance will be succeeded by the Sweat-lodge Ceremony. It is known to be a cleansing ceremony. The tribe will choose a particular spot to dig a fire-pit so rocks that have been specially chosen will be heated. The Sweat-lodge ceremony usually occurs in the late afternoon and could last until the dawn of the following day. This ceremony is known to be completed in two different ways. One way has heated rocks used and the other has water poured on the rocks. The main goal is to generate the effect of sweat. As the rocks are ready, the participants of the ceremony will remove their clothes where they are only wearing light undergarments or are naked. The ceremony host and participants will proceed to enter the lodge on hands and knees and then sit around the centre pit in a circle. The fire-tender of the ceremony passes the heated rocks around until they are placed into the pit. Various Indigenous tribes will have a different number of rocks as they will likely focus on a specific need for the community. The placement of rocks and their quantity is essential to what they are praying for. When the desired number passes into the lodge, the entry will be closed off and the ceremony's host will begin to pray. Following the prayers of the host and the participants, those in the lodge will leave the lodge to prevent any health hazards, and they proceed to repeat the ceremony, which can be up to four times.

These traditions continue to this day as tribes continue with the Sun Dance and Sweat Lodge ceremonies every year. Canada observes every June as National Indigenous History Month. During this month, Summer Solstice, which is known to be the day that it takes the sun the longest time till it sets, takes place in June. This has inspired the Summer Solstice Indigenous Festival. In a modern approach, this festival brings together and highlights Indigenous culture through music, art and celebrations.

SOLAR FOLKLORE

Storytelling is embedded in human nature across all cultures and tribes. It is what keeps the history of people alive throughout time. Storytelling is crucial for the Indigenous people as they have endured a long history of immense oppression and have been ordered to assimilate into the cultures of colonizers. That is why storytelling has an important role. It will preserve the culture of the Indigenous communities and will be passed on to the youth to keep the traditions alive and allows them to own a sense of belonging that will connect them to their roots. The sun plays an important role and has its meaning for different cultures. A common story known across the North American Indigenous tribes is 'Why There is Day and Night.' The story symbolizes how even within nature and animals there is an emphasis on the importance of the sun. The animals decided that time would be part 'day' and part 'night' with the help of the sun. It plays a role within their lives that allows them to see properly because the light of the sun and that part would be the 'day' so they hunt for food and gather resources to build their shelters. The animals also needed to rest and that part of the time when it is dark would be known as 'night.' This is relatively close to how most ancient tribes have lived whether they are Indigenous or not. The sun helps defines the day for those who seek it

THE IMPORTANCE OF SEASONAL CYCLES

As the sun heavily influences the cycle of the seasons, there are also traditions, celebrations and main rituals that are performed and practised throughout each season by Indigenous tribes. Some tribes even have a different set of seasonal cycles. For example, the Woodland Cree tribe follow a six-season cycle. The Woodland Cree tribe have their largest population located in Alberta, Canada. The tribe's six-season cycle consists of Winter, Spring, Break up, Summer, Fall and Freeze up. The two additional seasonals are meant to signify the importance of what is occurring in nature between Spring and Summer as well as Fall and Winter. The season between Fall and Winter known as the Freeze season is important to monitor for the people of the tribe, as the name of the season states, this is when the water of the region freezes up due to the cold weather. During the Freeze, travelling and moving around becomes more difficult because the temperatures are only cool enough to slightly freeze up the lakes. However, during the winter, the colder temperature will allow the tribes to be able to conduct their travels and activities without any danger. The Freeze helps identify how safe it is for people. As for the Break up, it is the same concept as the Freeze up, but it is when the lakes start to melt off, and once again it becomes unsafe for the people of the tribes to travel and conduct their activities once more. Since many people within the tribes depended on trade as it was necessary for their survival, they needed to have meaning for such periods as it would direct them away from what could be dangerous travels. As for the Winter, Spring, Summer and Fall seasons, specific activities would occur to maintain the Woodland Cree tribe's lifestyle.

SEASONAL TRADITIONS

An essential part of the solar calendar that is followed is the seasonals cycle that occurs throughout the year. Similar to the summer solstice, there is also a Winter Solstice that occurs during the season. Indigenous people place a lot of importance and honour towards seasonal changes as they are one of the most natural cycles on earth. An important tradition that takes place amongst many Indigenous tribes is the 'Round Dance.' The origin of the round dance also comes from storytelling and folklore. The Cree people believe that the Round Dance is assumed to have been started by a young girl who saw her mother's spirit, who went on to tell the young girl: "I cannot find peace in the other world so long as you grieve," she said, "I bring something from the other world to help the people grieve in a good way"(Cuthand, 2012). The mother's spirit goes on to teach her daughter a ceremony and songs so she can pass them on to the people. The ceremony should be the people celebrating their ancestors in a circle and she promises that the ancestors will be there to sing and dance with them and that is how it will allow the people to connect to their ancestors. The Cree believe that the Northern lights are the dancing spirits of the ancestors. The Northern Lights are famously known to appear during the winter seasons, which is why the Round Dance commonly occurs during the winter season.

SEASONAL IMPACT ON LIFE
Some Indigenous groups in Canada would surely have faced a different type of lifestyle due to the extremely cold weather in the geographic region where they lived. The Indigenous people of the Arctic that are also more commonly known as the Inuit, live mostly in the northern part of Canada. This Inuit tribe is heavily populated in Northwest Territories, Yukon, Nunavut, Labrador and Northern Quebec. The northern territories have recorded temperatures as low as 30 degrees Celsius. The Inuit tribes within these territories adapted accordingly to be able to survive these conditions. Clothing has been designed diligently for it to provide a way for the tribes in the North to survive such harsh winters they face. The clothing's material is meant to give humans the same advantage as the animals who inhabit the cold region. For instance, Caribous which are native to northern and Arctic regions have fur tightly packed hollow hairs meant to trap air and insulate the animal strongly. Clothing made from Caribou is one of the many materials used by animals along with material from beavers, moose, wolves, Dall sheep, snowshoe hares, muskrats, and marten. Together with these materials used to clothe the body and keep it warm, footgear is also of the utmost importance to have the correct material for. Sealskin is one of the materials used to make waterproof boots so they can be worn on sea ice or wet snow. The use of sealskin results in a pair of boots that is flexible and lightweight considering much of the clothing is heavier.

As a result of the extreme cold, much more innovation was needed to live throughout the winters such as the way the houses were built and the type of food that is eaten. However, it is important to understand that there is a lot of spirituality connected with the change of seasons and temperatures. Some tribes believe that the cold that occurs within the season change is a strong spirit that must be respected because if not, it may bring on alarming cold weather. Elders within tribes would become apprehensive when sudden drops in the weather occur as it often meant that someone is lying dead from the cold somewhere. Contrarily, the times we live in now bring in an awareness that the winters are not as extreme as before. " 'The weather is getting old,' some elders say because the cold is losing its strength and vitality." This way of thinking brings on an emerging issue that impacts not only the tribes that live in the North but the vast majority of Indigenous people, Climate Change.

Climate change is well defined as "the long-term shifts in temperatures and weather patterns. These shifts may be natural, through variations in the solar cycle." (United Nations, n.d.) The impact it has every single season is usually felt through the warmer temperature and that is why many people refer to it as "Global Warming." Indigenous people all over North America are facing difficulties due to life-threatening emergencies occurring due to climate change. Northern territories face a particularly difficult situation due to the melting glaciers. The Center for International Governance Innovation published an article describing one of the negative outcomes of climate change on Inuit people. "Glacial melt, long relied on for drinking water, is now unpredictable. In one stunning case, the Kaskawulsh Glacier in the Yukon has receded so far that its meltwater has changed direction, flowing south toward the Gulf of Alaska and the Pacific Ocean instead of north toward the Bering Sea. Ice that used to serve as our winter highways is giving way and invasive species are travelling much further north than ever before."(Watt-Cloutier, 2018) Additionally, the construction of pipelines on Indigenous lands is a human-driven action that causes unfavourable consequences. Most recently in 2022, the Indigenous people of the Wet'suwet'en lands in British Columbia were faced with "serious human rights violations as the construction of the Coastal GasLink pipeline has reportedly begun." (Amnesty International, 2022) The construction of this pipeline brings about the danger to the Indigenous people as it impacts the climate change issue which has been worsening as the seasons pass.

To conclude, the seasonal changes and the sun have had a great influence on the lives of Indigenous people from the very beginning of time. The influence flows through daily lives, traditions and celebrations. Preserving the stories and culture is essential to keep history alive for the generations to come as well as encouraging the protection of the land and heritage.

REFERENCES

Bial, R. (2015). The People and Culture of the Inuit. Cavendish Square Publishing, LLC.

Canada: Construction of pipeline on Indigenous territory endangers land defenders (2022, October 3). Amnesty International. https://www.amnesty.org/en/latest/news/2022/10/canada-pipeline-Indigenous-territory-endangers-land-defenders/

The climate in Northwest Territories Canada (n.d.). World Data. https://www.worlddata.info/america/canada/climate-northwest-

García, É. D., & Mazzetto, E. Mesoamerican Rituals and the Solar Cycle

Gaertner, D. (2015). Indigenous in cyberspace: CyberPowWow, God's Lake Narrows, and the contours of online Indigenous territory. American Indian Culture and Research Journal, 39(4), 55-78.

Gehrmann, V. (n.d.). ___ Chinese calendar - Chinese zodiac. Chinese Calendar – Chinese Customs. Retrieved March 15, 2023, from https://www.nationsonline.org/oneworld/Chinese_Customs/chinese_calendar.htm#:~:text=The%20solar%20calendar%20is%20a,from%2C%20as%20observed%20from%20Earth.

Lowie, R. H. (1915). The Sun Dance of the Crow Indians (Vol. 16). The Trustees.
Kidwell, C. S., Noley, H., & Tinker, G. E. (2020). A Native American Theology. Orbis Books.
O'Brien, S. C. (2020). Religion and Culture in Native America. Rowman & Littlefield Publishers.

McKenzie, L. (2017, August 2). Protecting the Body Fire. Encounters North. https://www.encountersnorth.org/cold-summary/2017/8/2/protecting-the-body-fire

Ogg, Arden. "Elder John Cuthand Shares the Story of the Round Dance." Cree Literacy Network, 26 May 2021, https://creeliteracy.org/2012/12/19/elder-john-cuthand-shares-the-story-of-the-round-dance/.

Šprajc, I., Inomata, T., & Aveni, A. F. (2023). Origins of Mesoamerican astronomy and calendar: Evidence from the Olmec and Maya regions. Science Advances, 9(1), eabq7675.

Watt-Cloutier, S. (2018). It's time to listen to the Inuit on climate change.

What Is Climate Change? (n.d.). United Nations. https://www.un.org/en/climatechange/what-is-climate-change

CHAPTER FOUR

NAVIGATION AND ORIENTATION

By Hafsah Saajidh

INTRODUCTION

Celestial navigation manifests itself in various ways throughout various Indigenous groups. From star maps, constellations, stories, and oral tradition, Indigenous use of the sky as a method of navigation takes many forms. While there are many similarities between Indigenous traditions' use of the sky, their diversity is represented in the multitude of ways the sky and its elements are used in navigation and orientation across various traditions. The sky and stars are not only used for travel, but also for timekeeping, calendars, preparing for seasons, and determining when to hunt. It is important to note, colonial forces not only encroached on Indigenous lands, but they also imposed their traditions, methods of navigation, and methods of knowledge-keeping on Indigenous peoples, as such, much of the traditional knowledge of Indigenous celestial navigation may have been lost. Despite this, Indigenous traditions have endured, and many still use traditional methods of celestial navigation to this day.

This chapter explores the navigational traditions of many Indigenous cultures. It touches on many cultures across North America and the Arctic, as well as Aboriginal and Torres Strait Islander peoples in Australia. This broad exploration provides insight into the deep-rooted and diverse traditions of Indigenous peoples. It also displays the plethora of uses and understanding of the sky and its celestial objects that play a central role in the navigation and orientation practices of many peoples.

SUB-ARCTIC INDIGENOUS GROUPS

NORTHERN DENE

Sub-Arctic Indigenous groups are made up of a multitude of ethnic groups including the Dene, Cree, Ojibwe, Atikamekw, Innu, Beothuk and Gwichyaa Gwich'in (Dene Nation, n.d). The Dene Nation is mainly located in the Northwest Territories and extends from present-day Alaska, across northern Canada, all the way to the southernmost tip of North America (Dene Nation, n.d). Celestial navigation is central to the Northern Dene. The geographic characteristics of this region are similar to the ocean, or the Arctic plains, as there is low legibility of landscape landmarks. Meaning that there are few distinctive and easily identifiable elements, which people can use to navigate the land, and identify their orientation (Cannon et al., 2022). Monotonous landscapes, that are flat and have few unique features, make it harder to memorize and navigate (Kelly, 2003). As such, the sky becomes an increasingly important element of navigation and orientation.

The sky plays an important role in timekeeping, especially when the sun may not come up at regular daily intervals. Dawn is recognized when The Big Dipper rises from the northeast horizon and points directly to the Arcturus star from mid-October to early December (Cannon et al., 2022). Time was also kept by tracking every moon that passed, as well as sunrises and sunsets when they were available. The orientation of certain stars and constellations was also kept track of to understand when daylight or nighttime would be approaching.

For the Yellowknife Northern Dene, star tracking plays a central role in travelling, especially during times when the barren landscape is covered in a whiteout (Cannon et al., 2022). Poles or sticks are placed on the ground at dawn to mark the southeast-northwest axis. When stars begin to appear in the sky after dusk, a unique and memorable star that aligns with the poles is chosen. The star is then followed, but because stars rotate through the sky, a new star has to be picked near the place of the original star at approximately 40-minute intervals. The new star that is picked is usually a little behind, or to the left of the star, as the movement of stars is typically on an east-to-west plane, and an angle to the horizon (Cannon et al., 2022). In this way understanding the stars are critical to being able to travel safely and keeping direction.

GWICHYAA GWICH'IN

The Gwichyaa Gwich'in reside in the Yukon Flats area of Alaska but also extend to the Northwest and Yukon territories of Canada (Gwichya Gwich'in First Nation, 2017). Their geographical area resembles and often overlaps with that of Yellowknife Northern Dene. Thus, similar characteristics of low legibility of landmarks are present. However, the Gwichyaa Gwich'in territory includes many dense forests and rivers, and their landscape shapes the unique ways in which they employ star navigation. The Gwichyaa Gwich'in methods of celestial navigation, differ from that of Yellowknife Northern Dene, in part due to the characteristics of their landscape. Compared to the Northern Dene which uses methods of star tracking for long-distance travelling in the open and barren country, the Gwichyaa Gwich'in employ methods for shorter trips, and often into thick forests (Cannon et al., 2022). They use a unique approach of a whole-sky constellation called Yahdii which resembles human anatomy.

As the constellation resembles the symmetry of human anatomy, the Gwichyaa Gwich'in can remember the spatial relationships between important stars. This constellation interacts with the landscape and rivers so that it can be used for orientation when views of landmarks and rivers are blocked by the thick forest (Cannon et al., 2022). The constellation of Yahdii is the spirit of a traveller who travelled around the world to make it a safer and more useful place for humans, by transforming it into its present form (Cannon and Herbert, 2020). Thus, the constellation is made up of 19 groups of stars that are mainly named after body parts, such as the right hand or the left hand of Yahdii (Cannon and Herbert, 2020). The constellation also helps with orienting the seasons. By observing the position of Yahdii, the Gwichyaa Gwich'in can determine how fast a season is approaching. For example, during fall, the left side of Yahdii falls below the evening horizon and becomes more centred in the morning. As it transitions from fall to winter, Yahdii becomes more centred in the evening (Cannon et al., 2022). Thus, the constellation is central to more than just travel.

GREAT LAKES

The Great Lakes regions are home to many Indigenous peoples. Including the Anishinaabeg, which includes the Ojibway, Odawa, and Potawatomi around the Great Lakes, Algonquian to the eastern woodlands, and Cree to the north and west of the woodland. The Anishinaabeg territory expands across a large area, extending from the great lakes to Quebec on the east, to the Rocky Mountains on the west, and down to Oklahoma on the south (Union of Ontario Indians, n.d). The Haudanausanee Confederacy is an alliance between six Indigenous nations including the Mohawk, Oneida, Onondaga, Cayuga, Seneca, and Tuscarora in many communities to the southeast of the Great Lakes (Haudenosaunee Confederacy, 2021). Just like the Northern Denee and Gwichyaa Gwich'in, the Anishinaabek and the Haudanausanee Confederacy, live in similar landscapes and within proximity to each other, however, their methods of celestial navigation differ. For example, the Haudanausanee Confederacy describes the area around the big dipper in terms of three hunters chasing a bear. In contrast, the Anishinaabeg describe it as 'Ojig' the Fisher and 'Maang' the Loon (Ontario Parks, 2022).

Anishinaabeg navigation depends on the north star, named 'Giiwedin Anag' which is derived from the verb "giiwe", which means to go home (Noodin, 2021). The name in itself depicts its importance to Anishinaabeg navigation. The sky and stars are described in relation to the land, portraying the Anishinaabeg belief in the relational dependence of life. For the Ojibwe, stars are understood in terms of the four seasons winter, spring, summer, and fall. There are specific constellations that appear more clearly during their corresponding seasons. This allows for maintaining a calendar and understanding orientation throughout the year (Ontario Parks, 2022). The fall constellation called 'Mooz' depicts a Moose (Ominika, 2020). It reflects the season of Moose hunting and signals that food and materials that are needed to last through the winter should start to be collected (Ontario Parks, 2022). The winter constellation of 'Biiboonkeonini', is the winter maker, which is the spirit that brings winter(Ominika, 2020). The spring constellation, 'Mishi Bizhiw', which is a panther (Ominika, 2020), warns of the thawing of ice which can cause seasonal flooding and danger. The panther is a symbol of this, as it lives in bodies of water and creates danger, by breaking the ice with its tail (Ontario Parks, 2022). The summer constellations of 'Ajiijaak', 'Nenabozho', and 'Noondeshin Bemaadizid'(Ominika, 2020), reflect the season of trapping and extra leisure time that are associated with summer (Ontario Parks, 2022). In this way, Anishinaabek, rely on the sky to prepare them for upcoming seasons and keep them in touch with the various phases of the year.

The Haudenausonee uses stars for orientation as well, especially to keep track of time throughout the day. The Morning Star brings the spirit of T'henden hawit'ha', who has the duty of transforming the night into day. This star signals a special time of day during which Haudenausonee are meant to send personal thanks to the Creator (Deyohahá:ge, 2006). The star cluster Pleiades also plays an important role not only in orientation but also in cultivation technology. The Pleiades has played a distinctive role in many diverse cultures throughout history. This is because the Pleiades is highly visible and have a distinct configuration (Ceci, 1978). The Pleiades start to become visible on the eastern horizon at the beginning of the fall season and are first viewed about an hour after sunset. On each succeeding night, they will be seen at a slightly higher position above the horizon, until they are directly overhead during midwinter. They begin to slowly descend towards the western horizon until they are no longer visible by spring (Ceci, 1978). Due to this, the Pleiades is a central star not only in predicting and orienting seasons but also to mark the time for specific agricultural practices.

MARITIMES: MI'KMAQ
The Mi'kmaq reside in eastern Canada and extend to the Northeastern area of the United States. They reside in the maritime provinces and the Gaspe peninsula of Quebec. They comprise seven districts, Kespukwitk, Sikepne'katik, Eski'kewaq, Unama'kik, Piktuk aqq Epekwitk, Sikniktewaq, and Kespe'kewaq (Cape Breton University, 2019). The Mi'kmaq join many Indigenous and non-Indigenous traditions that view the big dipper as a bear and heavily rely on it for their navigation and orientation. There are many accounts of Mi'kmaq star navigation particularly during the World Wars, as their unique position means that a large portion of Europe sits at the same latitude as Mi'kmaq communities (Bernard et al., 2015).

Stories also play a central role in Mi'kmaq sky navigation. The story of Muin and the Seven Bird Hunters is one of the most prevalent. It describes the movement of Ursa Major through a bear (Muin) and seven birds (Marshall, 2010). Muin provides the oral calendar from which the timing of Mi'kmaq celebrations and activities are calculated.

The positioning of Muin in the sky also determines when hunting is forbidden. During spring, she emerges, and hunting is not allowed because it is the time of breeding and when young are born (Harris et al., 2010). Muin moves across the horizon during summer with seven birds hunting her (Marshall, 2010) to signal that it is time to pick berries and begin activities in the local lakes without harming wildlife populations in the waters. In the fall, Muin rises on her hind legs, signalling the time for hunting and gathering for winter preparations (Harris et al., 2010). Muin is a tool for seasonal and activity orientation that is observed throughout the year.

INDIGENOUS PEOPLES IN AUSTRALIA

The Indigenous peoples in Australia are comprised of the Aboriginal and Torres Strait Islander peoples. They are two distinct groups, which are themselves composed of many groups that have distinct languages, cultures, and traditions (AIHW, 2021). These people have developed unique methods of navigation and techniques that have been passed down for many generations (Indigenous Knowledge Institute, n.d). It is important to maintain distinctiveness between these peoples, and avoid grouping them into a singular entity, as there is much diversity between them. Aboriginal is a broad term that is used to encompass the nations and custodians of Mainland Australia, as well as the Island of Tasmania, Fraser Island, Palm Island, Mornington Island, Groote Eylandt, and Bathurst and Melville Islands (Australian Government, n.d). Torres Strait Islanders refers to people from at least 274 small islands that span from the northern tip of Cape York in Queensland to the southwest coast of Papua New Guinea (Australian Government, n.d).

Concerning the sky, many Aboriginal and Torres Strait Islander traditions view the skyscape as a reflection of the landscape including rivers, forests, animals, and ancestral beings. For example, the milky way is often seen as a river, with the stars representing waterholes, mountains, and other natural landmarks (Indigenous Knowledge Institute, n.d). The celestial navigation methods and traditions of the Aboriginal and the Torres Strait Islander People, differ greatly from the Indigenous groups of North America that were previously discussed. This is because constellations and stars, such as the Polaris, that feature prominently in North American traditions, are not visible in the southern hemisphere. This means that cultures and traditions in the southern hemisphere developed unique navigation tools and techniques, that assist with navigation in their unique terrain and landscape. Many of the Indigenous populations in Australia take advantage of seasonal farming, and they kept track of the seasons by looking at the night sky. Their understanding of the night sky not only included the stars, but the sun, moon, and eclipses as well (Ridley, 2021). The Milky Way is central to many cultures across Australia and the Torres Strait Islands because it is visible directly overhead them (Bhathal, 2006). The Orion constellation, usually depicted as a young man or a group of young men, and the Pleiades often referred to as the Seven Sisters, also features prominently (Norris and Harney, 2014).

Much of the tradition is passed down verbally, thus celestial navigation includes a large oral component. This is evident in the use of songlines, which are creation songs passed down through generations and are named according to the language of the group. These songs contain maps of the land, and paths across the entire Australian continent (Norris and Harney, 2014). They also contain information about tides, eclipses, and the movement of celestial bodies throughout the sky, which allows for timekeeping, and maintaining a calendar. The songs are sung along a journey to remember landmarks, orientations, and constellations (EarthDate, 2020). Many of the modern highways in Australia align with these star maps and have been said to originate from Aboriginal songlines and maps (EarthDate, 2020).

One songline starts at Yirrkala in Arnhem island, which is where the Yolngu people believe Barnumbirr (Venus), crossed the coast as she brought the first humans to Australia. The song describes her path across the land and sky, including the locations of mountains and waterholes. The song contains an oral map, that is recognizable by many different languages along a path that stretches across the top of Australia (Norris and Harney, 2014).

THE WARDAMAN
The culture and language of the Wardman people are heavily influenced by astronomy. Their three major creation figures: Froglady Eathmother, Rainbow and Skyboss are all represented by dark clouds in the Milky Way. The Southern Cross constellation is central as it is used to

determine the Wardaman calendar and marks the cycle of dreaming stories throughout the year.

The Wardaman people heavily rely on celestial navigation, as most of the travel is done at night, so that the air would be cool, and the stars would be visible to use as guides. They also use the ecliptic, which is the path of the sun in the sky and the shadow of the moon as tools of navigation (Banks, 2018).

MABUAIG ISLAND
The Mabuaig Island lies in the Napoleon and Arnold Passages of the Torres Strait (Torres Strait Island Regional Council, 2016). The peoples on this Island have expert astronomers tasked with the responsibility of gathering information on the stars (The Australian Curriculum, n.d). When there is a star expected to appear, astronomers would rise early before daybreak to watch the sky and analyze the appearance and orientation of constellations. The collection of observations and information created a detailed knowledge base of stars and constellations that were used for navigation and orientation. The patterns were used to guide safe travels through the dangerous Strait. Due to the nature of navigation in the Strait, many constellations and star maps are universal across Islands and cultures. For example, the Baidam constellation and the Tagai constellation are essential to many traditions to orient North or South, by viewing their position along the horizon. The tagai is used by the Erubam Le to travel from Erub Island to Mer Island by steering to the left of the Tagai (The Australian Curriculum, n.d). Ultimately, celestial bodies including stars and constellations are central to the various cultures and lifestyles of the Aboriginal and Torres Strait Island Peoples.

CONCLUSION
Throughout this chapter, the richness and diversity of Indigenous cultures are fully displayed. One legacy of colonialism is that Indigenous culture, languages, and traditions have been grouped and portrayed as a pre-historic monolith. However, further exploration into Indigenous cultures proves that each Indigenous group has a unique identity, and way of life, for which they have developed tools for navigation and orientation that are of equal importance and use in the contemporary world as they were through many previous generations. This is

especially evident with groups such as the Northern Dene and Gwichyaa Gwich'in, who live in close territories, which in some cases overlap, yet use the same sky and stars for various purposes. The unique naming of constellations, and the various star stories between Indigenous groups, provide insight into the ways each group sees the world. It also shines a light on how different groups orient not only their days but also their weeks, months, and years. Despite this diversity and uniqueness, it is also interesting to note the points of interconnectedness and similarity between groups. It is interesting to note how their views and understandings of the sky differ from European and other Western understandings, and how this translates to differing lifestyles and worldviews. Ultimately understanding Indigenous celestial navigation not only displays the use of the sky in Indigenous traditions but also uncovers larger understandings of the various and diverse cultures.

REFERENCES

Australian Institute of Health and Welfare. (2021, September 16). Profile of Indigenous Australians. Australian Institute of Health and Welfare. Retrieved April 11, 2023, from https://www.aihw.gov.au/reports/australias-welfare/profile-of-indigenous-australians

Australian Government. (n.d.). Style manual. Aboriginal and Torres Strait Islander peoples. Retrieved April 11, 2023, from https://www.stylemanual.gov.au/accessible-and-inclusive-content/inclusive-language/aboriginal-and-torres-strait-islander-peoples

Banks, K. (2018, May 21). Aboriginal astronomy can teach us about the link between Sky and land | kirsten banks. The Guardian. Retrieved April 11, 2023, from https://www.theguardian.com/commentisfree/2018/may/21/aboriginal-astronomy-can-teach-us-about-the-link-between-sky-and-land

Bernard, T., Rosenmeier, L. M., & Farrell, S. L. (Eds.). (2015). Mi'kmawe'l Tan Teli-kina'muemk: Teaching about the Mi'kmaq. Eastern Woodland Print Communications.

Bhathal, R. (2006). Astronomy in aboriginal culture. Astronomy and Geophysics, 47(5). https://doi.org/10.1111/j.1468-4004.2006.47527.x

Cannon, C. M., Herbert, P., & Sangris, F. (2022). Yellowknives Dene and Gwich'in stellar wayfinding in large-scale subarctic landscapes. ARCTIC, 75(2), 180–197. https://doi.org/10.14430/arctic75292

Cannon, C., & Herbert, P. (2020). YAHDII IN THE NATIVE TRADITION: A

GWICH'IN (DENE) STAR CHART. Dinjii Zhuh Ky'àa Yahdii. Retrieved 2023, from https://uaf.edu/anlc/docs/Final%20Gwichin%20Poster%20-%20compresed%20for%20 web%20download%202-19-2020.pdf

Cape Breton University. (2019, October 3). The Mi'kmaq. Cape Breton University. Retrieved March 16, 2023, from https://www.cbu.ca/indigenous-affairs/mikmaq-resource-centre/the-mikmaq/

Ceci, L. (1978). Watchers of the Pleiades: Ethnoastronomy among native cultivators in northeastern North America. Ethnohistory, 25(4), 301. https://doi.org/10.2307/481683

Dene Nation. (n.d.). Home. Denendeh is the Land of the People. Retrieved March 16, 2023, from https://denenation.com/

Deyohahá:ge: (2006). Star knowledge. Retrieved March 17, 2023, from https:// snpolytechnic.com/sites/default/files/docs/resource/starknowledge.pdf

EarthDate. (2020, July 24). Aboriginal star maps. EarthDate. Retrieved April 11, 2023, from https://www.earthdate.org/episodes/aboriginal-star-maps

Gwichya Gwich'in First Nation. (2017, January 27). Gwichya Gwich'in first nation. Canada First Nations. Retrieved March 16, 2023, from https://www.first-nations.info/ gwichya-gwichin-nation.html#:~:text=Approximately%209%2C000%20Gwich'in%20 live,antiquity%20across%20Gwich'in%20lands.

Harris, P., Bartlett, C., Marshall, M., & Marshall, A. (2010). Mi'kmaq Night Sky Stories; Patterns of Interconnectedness, Vitality and Nourishment. CAPjournal, 9.

Haudenosaunee Confederacy. (2021, June 10). Who We Are. Haudenosaunee Confederacy. Retrieved March 16, 2023, from https://www.haudenosauneeconfederacy. com/who-we-are/

Indigenous Knowledge Institute. (n.d.). Stellar Navigation and mathematics. Stellar navigation and mathematics. Retrieved April 11, 2023, from https:// indigenousknowledge.unimelb.edu.au/curriculum/resources/stellar-navigation-and-mathematics

Marshall, L. (2010). Muin aqq l'uiknek te'sijik ntuksuinu'k: Mi'kmawey Tepkikewey Musikiskey a'Tukwaqn = muin and the seven bird hunters: A Mi'kmaw night sky story. (S. Kavanagh & K. Read, Eds.). Cape Breton University Press.

Noodin, M. (2021). wanitoon ani mikan odenang: Anishinaabe urban loss and reclamation. Urban History Review, 48(2), 16–31. https://doi.org/10.3138/uhr.48.2.02

Norris, R. p, & Harney, B. Y. (2014). Songlines and navigation in Wardaman and other Australian Aboriginal cultures. Journal of Astronomical History and Heritage, 17(2).

Ominika, W. (2020). Ojibwe astronomy. Great Lakes Guide. Retrieved March 16, 2023, from https://greatlakes.guide/ideas/ojibwe-astronomy

Ontario Parks. (2022, June 13). Stories in the stars/pride in our hearts. Parks Blog. Retrieved March 16, 2023, from https://www.ontarioparks.com/parksblog/indigenous-astronomy/

Ridley, A. (2021, June 16). Aboriginal astronomy. National Trust. Retrieved April 11, 2023, from https://www.nationaltrust.org.au/blog/aboriginal-astronomy/

Rockman, M., Steele, J., & Kelly, R. L. (2003). COLONIZATION OF NEW LAND BY HUNTER-GATHERERS. In Colonization of unfamiliar landscapes the archaeology of adaptation (pp. 44–59). essay, Routledge.

The Australian Curriculum. (n.d.). The Australian curriculum. Teacher background information (Version 8.4). Retrieved April 11, 2023, from https://australiancurriculum. edu.au/TeacherBackgroundInfo?id=16964#:~:text=The%20constellation%20of%20 Baidam%20(the the%20horizon%20to%20orient%20navigation.

Torres Strait Island Regional Council. (2016). Mabuiag. Go to the front page. Retrieved April 11, 2023, from https://www.tsirc.qld.gov.au/communities/mabuiag

Union of Ontario Indians. (n.d.). Who are the Anishinaabeg? UNION OF ONTARIO INDIANS. Retrieved March 16, 2023, from https://www.anishinabek.ca/education-resources/gdoo-sastamoo-kii-mi/who-are-the-anishinaabeg/

CHAPTER FIVE

INDIGENOUS ASTRONOMY AND SCIENCE

By Jeremy Steen

"We come from the stars, we are related to those stars. Once we finish doing what we came here to do, we go back up to those stars."
- Wilfred Buck

For centuries, Indigenous peoples around the world have faced colonization, forced assimilation, and systemic oppression; and yet, Indigenous societies, stories, and ways of living remain alive and resilient. Indigenous peoples represent descendants of the original inhabitants of a continent, such as North America or Australia. Indigenous peoples have systems of beliefs and practices that form the basis for their ways of life, just as any other society. One strong characteristic of Indigenous ways of living is a deep connection to nature and the universe. Astronomy is one way that Indigenous peoples have connected to the world; observations of patterns in the stars and sky inspired lore and life practices that have guided ways of life since the beginning of humankind. Indigenous knowledge of astronomy has been crucial in shaping the view of the sky today.

In this chapter, we will explore examples of astronomy in Indigenous societies around the world. We will especially consider how observations of the stars and sky inform traditions and practices that have helped Indigenous peoples survive and thrive throughout history. Next, we will examine differences between Indigenous and Western astronomy – the methods, motivations, and applications. We will also learn about how Indigenous astronomy was viewed by Western astronomers throughout history. Finally, we will look at collaborative efforts to promote Indigenous astronomy today.

THE SKY WOMAN – HAUDENOSAUNEE ORIGIN LEGEND

According to the Haudenosaunee, the world began in chaos and darkness, with no land and only animals in the dark. Then, the Sky Woman fell from a hole in the sky and landed on the back of a giant turtle. The animals of the water dove deep into the water to bring up mud from the bottom, and gradually they created the first land on the back of the turtle.

Sky Woman realized that she was pregnant, and she gave birth to a daughter who was named Tekawerahkwa. As Tekawerahkwa grew, she became curious about the world beyond their home on the turtle's back. So, Sky Woman sent her daughter on a journey to explore the new land. Tekawerahkwa flew out to explore the world, and she soon discovered that it was full of many different creatures, both good and bad. She saw that the world was beautiful but also dangerous. Along the way, she met several animals who became her helpers and guides, including the eagle, the beaver, and the bear.

When Tekawerahkwa returned home, she brought with her a bundle of seeds and gifts that she had received from her animal friends. Sky Woman used these gifts to create the world as we know it today. She planted the seeds, and they grew into plants and trees. She created rivers, mountains, and valleys, and she filled them with animals, birds, and fish. Thanks to Sky Woman and her daughter Tekawerahkwa, the world was transformed from darkness and chaos into a beautiful and diverse landscape. The Haudenosaunee people honour and celebrate their creation story to this day, as a reminder of the importance of respect for the earth and all its creatures (Niro et al., 2001).

ATIMA ATHCAKOSUK – THE CREE DOG STARS

The Cree call the seven stars sometimes considered the Little Dipper or Ursa Minor the Atima Atchakosuk – the Dog Stars. According to Cree lore, long ago, the people had no dogs. They had no companions for the children, no help for long journeys, and no warning system when danger was nearby. Relatives of the Cree, the Wolf, Fox, and Coyote, decided to send some of their young to live with the people and to protect them. Two wolf pups, two fox pups, and two coyote pups went to live with humankind. They adapted into the dogs that now inhabit our world. They guard our homes, our communities, and our loved ones.

To honour the sacrifice made by the Wolf, the Coyote, and the Fox, the Creator created Atima Atchokosuk, a constellation that represents a reminder of the dogs in the heavens. The three dogs guard the people against the sky, anchored by a leash to Polaris, the North Star (Boutsalis, 2021).

THE STORY OF THE EMU – TORRES STRAIT ISLANDERS ORIGIN STORY

According to the lore of the Australian Aboriginals of the Torres Strait Islands, emus are more than just birds – they are the creator spirits that soared through the skies, guarding over the people below. The Emu can be seen in the sky. It is represented by what others might call the Milky Way. The dark between the stars makes up the outline of the Emu. Depending on the time of year, the Emu is seen either sitting or running, and this positioning tells Indigenous peoples whether they should be hunting for emus or collecting emu eggs (Hamacher, 2014).

THE IMPORTANCE OF ASTRONOMY IN INDIGENOUS SOCIETIES

From the examples above, we can see that the stars and the sky feature heavily in stories of creation and tradition in Indigenous societies. Further to its importance in spirituality, astronomy also plays an important role in the practical ways of Indigenous peoples. In the Story of the Emu, we saw a glimpse of how the presentation of the Milky Way galaxy offered cues to Indigenous peoples of the Torres Strait Islands about the hunting and gathering seasons. In what is now known as North America, the Anishinaabeg people use the 13 moon cycles to track the seasons. Each moon cycle is named according to important practices or features of that time of year (Center for Native American Studies, n.d.; Kanawayhitowin, n.d.): for example, the Miin Giizis (Berry Moon) in July, berry-picking season; the Minoomini Giizis (Grain/Wild Rice Moon) in August, rice-picking season; and Minado Giisoonhs (Little Spirit Moon), which is manifested in the Aurora Borealis (Northern Lights). Another example of Indigenous peoples using astronomy for practical purposes is the Torres Strait Islanders' use of the scintillation (twinkling) of stars to tell when a monsoon is arriving (Signorelli, 2019). The same peoples of the Torres Strait Islands also use the location and direction of the sunset along the horizon to determine the changing of the seasons and animal behaviours (Hamacher et al., 2020a).

DIFFERENCES BETWEEN INDIGENOUS AND WESTERN ASTRONOMY

Indigenous astronomy and Western astronomy have developed in vastly different ways throughout human history. Indigenous astronomy is rooted in the observations and interpretations of the natural world by various Indigenous cultures, while Western astronomy is based on a more scientific, observational approach to understanding the universe (Signorelli, 2019).

One of the most significant differences between Indigenous and Western astronomy is how each culture approaches the study of the stars and other celestial bodies. Indigenous cultures have long relied on oral tradition and storytelling to pass down knowledge about astronomy from one generation to the next. This knowledge is often tied to the land and is intimately connected to the natural world around them. Indigenous people often use astronomy to track seasonal changes, predict weather patterns, and determine the best times to plant and harvest crops.

Western astronomy, on the other hand, is based on the scientific method, which involves careful observation, experimentation, and data analysis. Western astronomers use telescopes, satellites, and other instruments to study the universe and make predictions about its behaviour. They seek to understand the fundamental laws of physics and the workings of the universe at a deeper level.

Another key difference between Indigenous and Western astronomy is how each culture approaches the spiritual and cultural significance of the stars. For many Indigenous cultures, the stars are seen as sacred beings with their personalities and relationships with one another. They are often tied to creation stories and the history of the people and are used to navigate both physical and spiritual journeys (Hamacher et al., 2020a). In contrast, Western astronomy tends to view the stars and other celestial bodies as objects to be studied and analyzed. While some Western cultures may attach symbolic or cultural meaning to certain stars or constellations, these interpretations are not typically central to the study of astronomy itself.

There are also differences in the types of knowledge and skills required for Indigenous and Western astronomers. Indigenous astronomers often rely on their knowledge of the land and the natural world to make observations and predictions about the stars. This requires a deep understanding of the environment, as well as a keen sense of observation and intuition. In the Western tradition, astronomers require extensive training in mathematics, physics, and other sciences to fully understand the workings of the universe. They must also be skilled in the use of sophisticated instruments and technology, such as telescopes and computers, to make precise observations and measurements (American Astronomical Society, n.d.).

SIMILARITIES BETWEEN INDIGENOUS AND WESTERN ASTRONOMY

Despite these differences, there are also many similarities between Indigenous and Western astronomy. Both have developed sophisticated methods for observing the stars and other celestial bodies, including tracking their movements over time. Both also emphasize the understanding of how the stars and the sky are related to life on Earth – the seasons, the tides, and animal behaviours, to name a few.

Recognition of Indigenous Astronomy by Western Science

The recognition of Indigenous astronomy by Western scientists has been a slow and often fraught process. For centuries, Western astronomers dismissed Indigenous knowledge as superstitious or unscientific, failing to recognize the valuable insights these cultures had to offer. However, in recent years, there has been growing recognition of the value of Indigenous astronomy within the scientific community, leading to a renewed appreciation for the contributions of Indigenous astronomers throughout history.

One of the earliest examples of Western recognition of Indigenous astronomy comes from the work of Spanish Jesuit missionaries in the 16th century. These missionaries travelled to the New World intending to convert Indigenous peoples to Christianity, but they also recorded detailed observations of Indigenous astronomical practices. They were particularly struck by the precision with which Indigenous astronomers were able to predict astronomical events, such as eclipses, and recognized the sophistication of their astronomical knowledge (Harris, 2005).

Despite this revelation, in the centuries that followed, Western astronomers largely ignored Indigenous knowledge, seeing it as inferior to their scientific methods. However, in the 20th century, there were some notable exceptions.

RAY NORRIS

In the 1960s and 70s, the Australian astronomer Ray Norris worked closely with Indigenous communities in Australia to document their astronomical knowledge. He recognized the deep connections between Indigenous cultures and the natural world and saw the importance of preserving this knowledge for future generations.

Norris shared his learnings about Indigenous astronomy during his tenure in the Department of Indigenous Studies at Macquarie University in Sydney, Australia (Our World, n.d.). He also published a book about Indigenous astronomy called Emu Dreaming, with his wife Cilla Norris (2009).

WILFRED BUCK

Similarly, in the 1990s, Canadian astronomer Wilfred Buck began working with Indigenous communities in Canada to promote the study of Indigenous astronomy. He saw the value of combining Western scientific methods with Indigenous knowledge and worked to bridge the gap between these two approaches. His work has since inspired a new generation of Indigenous astronomers, who are working to reclaim and revitalize their traditional astronomical practices. Buck has shared his knowledge at schools and universities across Canada and is known for guiding students and youth through the difficult conversations surrounding colonialism, oppression, and Indigenous knowledge (Office of the Provost and Vice-President, Carleton University, 2022).

DEVELOPMENTS IN THE 21ST CENTURY

In recent years, there has been a growing recognition of the value of Indigenous astronomy within the scientific community. This has been driven in part by the recognition of the limitations of Western scientific methods, particularly when it comes to understanding the complex relationships between the natural world and the cosmos. Indigenous knowledge can offer valuable insights into these relationships and can help to fill gaps in Western scientific knowledge. Despite these recent developments, there is still much work to be done to fully recognize and appreciate the value of Indigenous astronomy. Indigenous knowledge has often been marginalized or ignored by Western scientific institutions, and there is a need for greater collaboration and recognition of the contributions of Indigenous astronomers throughout history.

Overall, the recognition of Indigenous astronomy by Western scientists throughout history has been a slow and often fraught process. However, as we begin to appreciate the value of Indigenous knowledge in the study of the cosmos, there is hope that we can build a more inclusive and collaborative approach to understanding the universe and our place within it.

WHY DO WESTERN SCIENTISTS NEGLECT INDIGENOUS ACHIEVEMENTS IN ASTRONOMY?

The Australian astronomer Ray Norris spent many years researching Indigenous astronomy in Australia, and he wrote and spoke extensively about what he found. He also wrote about his theory for why Western scientists have historically been so reluctant to recognize Indigenous astronomy and other Indigenous achievements. Norris referred to something called 'The Paradigm Problem' – the idea that many Westerners, as a result of colonialism and oppression, grew up with the notion that Indigenous methods are inferior; this makes it difficult to accept information outside of that paradigm (belief; Norris, 2014). Not long ago, Indigenous peoples around the world were still described as primitive or savage. Though new understanding has begun to change that idea, the underlying belief is slow to catch up.

COLLABORATIVE EFFORTS TO PROMOTE INDIGENOUS ASTRONOMY

Collaborative efforts to promote Indigenous astronomy research have been growing in recent years, as scientists recognize the value of integrating Indigenous knowledge into their work. These efforts involve partnerships between Indigenous communities and scientific institutions and aim to promote mutual understanding and respect while advancing our collective understanding of the cosmos. Let's explore some examples of these partnerships around the world:

In Australia, the Australian Indigenous Astronomy (AIA) group shares both traditional knowledge and modern research about astronomy. They also played a large role in designing curriculum elements for schools to teach children about Indigenous astronomy and ways of living (Aboriginal Astronomy, 2022). Researchers from the AIA group also aim to raise awareness about threats to traditional Indigenous practices posed by modernization. One example is light pollution, or the electric light is given off by major cities that makes it difficult (sometimes impossible) to observe the stars (Hamacher et al., 2020b).

In Canada, the Canadian Astronomical Society (CASCA) has established a working group dedicated to promoting Indigenous astronomy research. This group called the Long Range Plan Community Recommendations Implementation Committee (LCRIC), works closely with Indigenous communities and organizations to build relationships and develop research projects that integrate Indigenous knowledge and perspectives. The LCRIC hosts discussion groups, delivers educational workshops, and overall aspires to increase transparency and inclusion in the field of astronomy (Morsink, 2022). Another example in Canada is the Astronomy in Indigenous Communities (AIC) project, which aims to use astronomy as a developmental tool for Indigenous communities by providing opportunities to explore the stars and the sky and learn about astronomy from both Indigenous and Western perspectives (Moumen et al., 2020). To highlight one further example of Canadian efforts to promote Indigenous astronomy research, Wilfred Buck, the Canadian astronomer discussed earlier in the chapter, hosts weekend events for students to learn about Indigenous astronomy called 'Tipis and Telescopes' (Kirby-McGregor, 2021).

In the United States, artist Annette Lee founded the Native Skywatchers as a research and programming initiative designed to communicate Indigenous knowledge surrounding sustainable living and harmony with the sky and the earth (Native Skywatchers, 2007).

In all of the above cases, these collaborative efforts recognize the value of Indigenous knowledge and the importance of respecting and incorporating it into scientific research. They also aim to empower Indigenous communities and promote self-determination, recognizing that Indigenous peoples are the best stewards of their knowledge and culture.

One key aspect of these collaborative efforts is the need for respectful engagement and dialogue between Indigenous and non-Indigenous partners. This requires a deep understanding of the cultural and historical context in which Indigenous knowledge is situated, as well as a willingness to listen and learn from Indigenous perspectives.

Another key aspect is the need for ethical considerations when conducting research in Indigenous communities. This includes obtaining free, prior, and informed consent from community members, respecting community protocols and intellectual property rights, and ensuring that research benefits Indigenous communities in meaningful ways (Panel on Research Ethics, 2022).

CONCLUSION

In this chapter, we reviewed examples of Indigenous culture and ways of living manifested in astronomy. We also discussed the tumultuous history of relations between Indigenous astronomy and Western scientists, as well as recent efforts to ameliorate the relationship and promote Indigenous astronomy research. Furthermore, collaboration can address historical injustices and help bridge the gap between Indigenous and non-Indigenous communities. Western astronomy has a history of colonization and exclusion of Indigenous peoples, and collaborating with Indigenous astronomers can help address these issues and create a more equitable scientific community.

Indigenous astronomy also has the potential to inspire new research questions and directions for Western astronomy. By working together, Indigenous and Western astronomers can create a more comprehensive understanding of the universe and develop new technologies and techniques for observing and studying it. A collaboration between Indigenous and Western astronomers is crucial for advancing the field of astronomy, promoting diversity and inclusion in science, addressing historical injustices, preserving Indigenous cultures and knowledge, and inspiring new research directions. It is important to recognize the value of Indigenous knowledge systems and to work towards a more collaborative and inclusive scientific community.

REFERENCES

Aboriginal Astronomy (2022). Australian Indigenous Astronomy. Retrieved from http://www.aboriginalastronomy.com.au/.

American Astronomical Society (n.d.). What does it take to become an astronomer? Retrieved from https://aas.org/faq/what-does-it-take-become-astronomer#:~:text=Most%20research%20astronomers%20have%20doctorateto%20become%0a%20research%2astronomer.

Boutsalis, K. (2021). Teaching Indigenous Star Stories. The Walrus. Retrieved from https://thewalrus.ca/space-teaching-indigenous-star-stories/?gclid=Cj0KCQjwwtWgBhDhARIsAEMcxeB_ZmiwPY369dIJFGkSgJg0galT98YMQvAKCQAr4OBwKdjOEx9DbgaAr_nEALw_wcB

Centre for Native American Studies (n.d.). Moons of the Anishinaabeg. Retrieved from https://nmu.edu/nativeamericanstudies/moons-anishinaabeg-0#text=The%20Anishinaabeg%20people%20live%20in,influence%20within%20a%20given%20location.

Hamacher, D. W. (2014). 'Dancing with the stars': Astronomy and Music in the Torres Strait. Retrieved from http://www.aboriginalastronomy.com.au/wp-content/uploads/018/05/Hamacher-Dancing-with-Stars.pdf

Hamacher, D. W., Fuller, R. S., Leaman, T. M., & Bosun, D. (2020a). Solstice and solar position observations in Australian Aboriginal and Torres Strait Islander traditions. Journal of Astronomical History and Heritage, 23(1). https://doi.org/10.48550/arxiv.2001.08884

Hamacher, D. W., de Napoli, K., & Mott, B. (2020b). Whitening the Sky: light pollution as a Form of cultural genocide. Journal of Dark Sky Studies, 1. https://doi.org/1048550/arxiv.2001.11527

Harris, S. J. (2005). Jesuit scientific activity in the overseas missions, 1540-1773. Isis, 96(1), 71-79. https://doi.org/101086/430680

Kanawayhitowin. (n.d.). Moon Teachings. Retrieved from http://www.kanawayhitowin.ca/pageid=214

Kirby-McGregor, L. (2021). Star stories: The making of the Indigenous Star Knowledge Project. Ingenium Channel. Retrieved from https://ingeniumcanada.org/channel/articles/star-stories-the-making-of-theindigenous-star-knowledgeproject.

Native Skywatchers (2007). About Native Skywatchers. Retrieved from https://www.nativeskywatchers.comabouthtml.

Niro, S., George, K., & Brant, A. (2001). Origin Stories – Sky Woman. Retrieved from https://www.historymuseum.ca/cmc/exhibitions/aborig/fp/fz2f22e.html.

Norris, R. (2014). Aboriginal people – how to misunderstand their science. The Conversation. Retrieved from https://theconversation.com/aboriginal-people-how-to-misunderstand-their-science-2835

Norris, R. & Norris, C. (2009). Emu Dreaming.
Moumen, I., Rousseau-Neption, L., Cowan, N., Safi-harb, S., Bolduc-Duval, J., & Laychak, M. B. (2020). Astronomy in Indigenous Communities. Communicating Astronomy with the Public Journal, 27, 27-30.

Morsink, S. (2022). Report from the LCRIC. Cassiopeia, 2022. Retrieved from https://casca.ca/cat=11

Signorelli, L. (2019). What is Indigenous astronomy? We spoke to two experts to find out. Australian National Maritime Museum. Retrieved from https://www.sea.museum/2019/08/09/what-is-indigenous-astronomy-we-spoke-to-two-experts-to-find-out#:text=In%20astronomy%2C%20Indigenous%20people%20had,over%20long%20periods%20of%20time.

Our World (n.d.). Ray Norris. United Nations University. Retrieved from https://ourworld.unu.edu/en/contributors/ray-noris

Office of the Provost and Vice-President, Carleton University (2022). Indigenous star-lore expert Wilfred Buck dazzles Carleton community with astronomical insights. Carleton University. Retrieved from https://carletonca/provost/2022/10/indigenous-star-lore-expert-wilfred-buck-dazzles-carleton-community-with-astronomical-insights

Panel on Research Ethics (2022). Tri-Council Policy Statement: Ethical Conduct for Research Involving Humans. Government of Canada. Retrieved from https://ethics.gc.ca/eng/policy-politique_tcp2-eptc2_2022.html

CHAPTER SIX

WESTERN KNOWLEDGE AND TRADITIONAL KNOWLEDGE

By Odette Will

SHADES OF KNOWLEDGE

Every field and generation has contributed to the society that we know today. There must be a brilliant mind behind traditional knowledge of astronomy and its similarities to the knowledge we know today. One can argue that traditional knowledge of astronomy is not based on science and that it is mere storytelling. But this is simply not true. A recently published paper accounted for the experience of Finnish scientists. Initially, the scientists did not believe that bright auroral displays made audible sounds. The Finnish scientists conducted an experiment using Sami traditions and heard the light phenomenon's sounds (Laine, 2022). As we can see from this story, traditional knowledge is not always considered, but all knowledge should be treated with equal importance.

I, the author of this chapter, have been challenged to do research outside my discipline. For the first time, I embarked on a journey of learning about astronomy and traditional knowledge. Honestly, I had to research the importance of astronomy to understand the topic better.

To begin, astronomy is deeply rooted in the history of almost every society due to its practical applications and philosophical implications (Percy, 1998). It still has everyday applications to timekeeping, seasons, navigation and climate, and longer-term issues such as climate change and biological evolution (Percy, 1998). Astronomy contributes to the development of physics and other sciences (Percy, 1998). It shows our place in time and space and our kinship with other peoples and species on Earth (Percy, 1998). There is a concept known as One Health. One Health is an approach that recognizes that the health of people is closely connected to the health of animals and our shared environment. This has been a hot subject in recent years, yet the idea is very traditional. This shows that we apply preserved traditional knowledge without knowing its origins. Indigenous people have always believed that all living and nonliving things and natural and social worlds are intrinsically linked (Toledo, 1999).

On to my next question, how do people benefit from astronomical science? In population health, science, we use our research to show that many factors contribute to health outcomes, and we can all argue that health is important. But, if it was not for astronomy there would not have been Wi-Fi-meaning no GPS, or social media. Astronomy has a unique ability to unite humans (How Can Astronomy Improve Life on Earth? | Center for Astrophysics, n.d.). By simply asking big questions about the Universe and our place in it, we see ourselves as we are: together, voyaging through a singular moment in time on one very special but relatively minuscule planet among the vastness of space (How Can Astronomy Improve Life on Earth? | Center for Astrophysics, n.d.). If this is a science that can bring unity, then it should bring unity to all people. My perspective is that some people have been left out of this science. In the past, we have seen men dominate this workforce, and today, I cannot name an Indigenous person in this field of work. For there to be unity or collaboration, there needs to be an adoption of a new framework or knowledge, in this paper, I will argue that this new knowledge is the old knowledge that has been utilized.

The purpose of this chapter is to extract the body of traditional knowledge of astronomy of different Indigenous ethnic groups; to understand the ways and degrees through which this knowledge and beliefs shaped the lived realities of Indigenous people; and to add to our understanding of Indigenous scientific practices, and this can be used to complement science education.

The Importance of Preserving Traditional Knowledge of Astronomy
I have researched Indigenous health, and little did I know that preserving traditional knowledge is key to restoring and maintaining the health and well-being of Indigenous communities worldwide. There is a need for Indigenous knowledge in other fields of study as well. In this paper, I

will use the terms Traditional knowledge and Indigenous knowledge interchangeably.

Traditional knowledge refers to indigenous peoples' knowledge, innovations and practices. Developed from experience gained over the centuries and adapted to the local culture and environment, traditional knowledge is often transmitted orally from generation to generation. It tends to be collectively owned and can be expressed in stories, songs, folklore, proverbs, cultural values, beliefs, rituals etc. (Secretariat on the Convention on Biological Diversity, n.d.b). It can also be called Indigenous knowledge, and in this chapter, I will sometimes abbreviate TK or IK.

Although an increasing number of Indigenous people live less traditional lives, many still seek to maintain meaningful connections to their traditional knowledge, heritage, and land (Edwards & Heinrich, 2006). Such priorities for continuing their traditions and transmitting Indigenous knowledge to future generations hold a special place in the United Nations Declaration on the Rights of Indigenous Peoples (United Nations, 2008). Indigenous people also have different ways of managing access to, and use of, knowledge which are differentiated at three general levels: public areas (open-access), peri-restricted areas (requiring negotiation for access and use terms); and highly restricted or closed areas (secret-sacred knowledge sites, practices and documentation) (Gumbula, 2005). Today, there are efforts in technology to document traditional knowledge. But why? Let us use the social media Tik Tok as an example. On Tik Tok you can do more than just capture videos, you can share the videos too. Modern technologies are used in hopes of sharing knowledge. It is easier to keep and refer to electronic copies than paper copies. With modern technology, the hope is to preserve knowledge and not lose it like a sheet of paper. Preserving traditional knowledge can empower the younger generation of Indigenous people and help maintain the diversity that our world needs. This is crucial because as elders in the communities pass away knowledge can be lost (that is if the knowledge was not documented). The knowledge can be used for the current and future generations to come.

If we want to see unity, then it is important to maintain traditional knowledge of astronomy. We have seen division between society and Indigenous people. If we can maintain the traditional knowledge, then we can use it to involve Indigenous communities in astronomical research. Traditional knowledge can promote self-identity, self-reliance (especially the ability to support traditional lifestyles) and self-government by creating a strong, ongoing appreciation within the community of its history and its roots (Government of Canada, Public Services and Procurement Canada, Integrated Services Branch, Government Information Services, Publishing and Depository Services, n.d.). Preserving a community's IK is different from protecting it from misuse by others. Communities may want to preserve their knowledge for several reasons. Some communities have identified a range of economic benefits to be gained from sharing their IK with others (Government of Canada, Public Services and Procurement Canada, Integrated Services Branch, Government Information Services, Publishing and Depository Services, n.d.).

THE CHALLENGES AND OPPORTUNITIES FOR SHARING TRADITIONAL KNOWLEDGE

Indigenous knowledge is usually presented orally. Some challenges include communication barriers arising from the different languages and styles of expression used by traditional knowledge holders; conceptual barriers, stemming from the organizations' difficulties in comprehending the values, practices, and context underlying traditional knowledge; political barriers, resulting from an unwillingness to acknowledge traditional-knowledge messages that may conflict with the agendas of government (Ellis, 2005). This requires researchers also know this area, ensuring collaboration with relevant Elders and communities. It is also important to note that some communities will refuse participation as star knowledge is subject to particular restrictions, and public or outsider access will not be possible (Hamacher & Nakata, 2023). Some Indigenous people think that their traditional knowledge is not valued as a useful contribution to environmental practices, land resource management, or broader scientific knowledge (Nakata et al, 2014).

In many places, this knowledge has been fragmented and Elders with a deep knowledge have passed away without that knowledge being passed down to younger generations (Hamacher & Nakata, 2023). Still, other barriers emanate from the co-opting of traditional knowledge by non-aboriginal researchers and their institutions. These barriers help maintain a power imbalance between the practitioners of science European-style environmental governance and the Aboriginal people and their traditional knowledge (Ellis, 2005). This imbalance fosters the rejection of traditional knowledge or its transformation and assimilation into Euro-Canadian ways of knowing (Ellis, 2005). Although this source was dated to the early 2000s, we still need to keep this in mind today. The same is true for other disciplines that want to study Indigenous People. In health research, we talk all about trust building, and that trust building can take years.

It takes an open mind to learn about traditional knowledge. For so long, European worldviews have been superior, universal, normative, and ideal, which is part of maintaining Aboriginal identity, language, and culture in modern society (Battiste M., 2000, pp. 192-193). In some cases, traditional knowledge has been abused. I think that for too long educators and researchers were talking about Indigenous people instead of talking with them. Society has isolated Indigenous people, so building trust takes a long time. If we are seeking ways how to translate Indigenous knowledge, then we also need to think about how we are supporting the communities. We also need to figure out how to be better communicators. The right message needs to go to the right population and at the right time. There needs to be a collaborative approach with community members and academia. Unfortunately, few academics are familiar with this subject.

Common research topics often seek to answer urgent questions that scientists and society need to know. We saw this with Covid-19, the rush of scientists to conduct studies on this topic. The same is true with traditional knowledge. The study of Indigenous astronomy - the bodies of locally developed knowledge about the stars - is an increasingly popular topic globally (Hamacher & Nakata, 2023). One of the driving forces behind this research is to confront and correct ongoing colonial practices that degrade traditional Indigenous ways of knowing by dismissing them as "myth and legend" (Clements et al, 2021). This emerging research topic has sparked a new challenge for researchers, and this can be an opportunity to help Indigenous communities. This is more important than a researcher focusing on advancing their career. This is not only about benefiting minority Indigenous groups, this is about benefiting society.

Where there is a gap in the literature, there is an opportunity for academic research. At my institution, students are required to take Indigenous studies and graduate students can take a truth and reconciliation course. More students want to study Indigenous perspectives and apply what they learn in their thesis. This is evidence that the minds of the younger generation are eager to understand a different perspective. I used to think that Indigenous knowledge is not backed up by science, but I know now that I am wrong. In Saskatchewan, there are schools for First Nations students, so it is potential for the younger generation to learn about traditional knowledge.

ROLE OF INDIGENOUS EDUCATION IN PROMOTING TRADITIONAL KNOWLEDGE

Today, we would like to incorporate Indigenous knowledge in research, which is not only true in astronomy. I have already mentioned the barriers to Indigenous research. One way of overcoming the barriers is by using education as a tool to promote traditional knowledge. I, the author of this paper, have already shared my thoughts on the topic; initially, I did not see the value of traditional knowledge applied to astronomy. When I was in school, I was taught in such a way that it was storytelling. There needs to be an integration of Indigenous perspectives in school curriculums because for so long school children were not told the full truth. I only happened to start learning about this in recent years. We need an accurate representation of IK. Educators (not only researchers) can include Indigenous knowledge and voices. Indigenous astronomy knowledge has to be collaboratively developed with tribes so accurate information can be shared (S Lee et al., 2020). Promoting IK is promoting diversity in education.

I have already identified that a challenge in sharing traditional knowledge is partially explained by youth not being in touch with their culture. Integration of Indigenous culture into curriculums will teach Indigenous youth about their culture. This can lead to empowering the Indigenous youth. The Truth and Reconciliation Commission of Canada published its calls to action in 2015 with 94 recommendations. Many of these 94 recommendations are directly related to education, language, and culture, some of which the Canadian Astronomy community can address and contribute to as part of reconciliation. If Canadian school systems are to move in the direction of developing improved modalities of science education involving traditional environmental knowledge, there are significant areas of caution that educators need to be cognizant of in extracting any type of traditional knowledge from Aboriginal communities (Vizina, 2010). Currently, there is little to no guidance for educators concerning these issues, or issues related to Indigenous community protocols, which can be even more complex and diverse (Vizina, 2010). Without the basis of ancestral knowledge, the contemporary colonial school curriculum offering only a Eurocentric science curriculum effectively ends the cycle of perpetuating traditional knowledge by continuing the process of assimilation in Canada (Vizina, 2010). This is not only a tremendous loss to Indigenous populations but to global collaboration on the conservation of biodiversity.

Community-based cultural education and traditional learning programs can be identified as an integral part to preserve traditional knowledge of astronomy. The role of Indigenous education can lead to the creation of regional-based Indigenous knowledge exchanges that could bring together youth and Elders from neighbouring communities (Protecting and Preserving Traditional Knowledge and Culture - Indigenous Climate Hub, 2018). This further leads to mentorship for youth. Ultimately, the aim of this work is increasing STEM/Informal Science Learning opportunities for Indigenous youth, adults, and communities, but also has the foundational goals of increased cultural pride, engagement in science, and community wellness. For the non-Native audience, there is great value in learning and practising cultural agility. Standing together we have enormous reach and capacity (S Lee et al., 2020).

STRENGTHS, LIMITATIONS OF THE STUDY AND FURTHER SCOPE

Although I did not write this article with Indigenous people or sought guidance from an Indigenous person, this article responds to a need proposed by them. This paper revealed that there is value in traditional knowledge, and it is not mere storytelling. This paper is a contribution towards a new direction in science which is moving away from colonialism. A researcher from any scientific discipline can learn from this article, and expand on the concepts presented in this chapter.

When I mentioned Indigenous people, I was talking about Indigenous people at the national level, but most of the articles were based in Australia. Some of the articles mentioned were very old. Even though I used credible articles most of them were dated back to the late 90s/ early 2000s and those may not capture the current knowledge that is known today. I also was not able to get full access to some articles because it was not available through my institution, so I could have included more information for this chapter.

In the future, research should be done on the ways educators can integrate traditional knowledge into the education systems. Further studies should also mention ways to overcome the challenges. I mostly mention what challenges are present and that it requires patience to build that trust. Lastly, I did not mention strategies for building trust or how to collaborate with Indigenous communities. Therefore, I urge readers to read articles on collaborative action-based research.

CONCLUSION
Every generation and culture has made a contribution that can be used today. There were brilliant minds behind TK, and now those who study astronomy are trying to capture this knowledge. Indigenous knowledge is holistic and it can give us the bigger picture, and it uses observations and experiences, and you can argue that it is scientific instead of myths.

In this chapter, I explored three main topics: the importance of keeping TK, the opportunities and challenges of TK, and the role of Indigenous education in promoting TK. While researching, I had an open mind and I learned a few things. The research was a challenge, but I think a greater challenge is ahead. There needs to be an emphasis on education that promotes TK because that is the way that we can implement the research. Some of the points that I raised in this paper do not only apply to astronomy, but to other sciences as well. I urge readers in other sciences to implement some of the things they have learned in this book or to use this book as a reason to focus on Indigenous research.

REFERENCES

Ellis, S. C. (2005). Meaningful Consideration? A Review of Traditional Knowledge in Environmental Decision Making. Arctic, 58(1), 66–77. http://www.jstor.org/stable/40512668

Government of Canada, Public Services and Procurement Canada, Integrated Services Branch, Government Information Services, Publishing and Depository Services. (n.d.). Information archivée dans le Web. Information Archived on the Web. https://publications.gc.ca/collections/Collection/R2-160-2001E.pdf

Gumbula, J. N. 2005. "Exploring the Gupapuynga Legacy: Strategies for Developing the
Galiwin'Ku Indigenous Knowledge Centre." Australian Academic & Research Libraries, 36 (2):23–26.

How can astronomy improve life on Earth? | Center for Astrophysics. (n.d.). https://www.cfa.harvard.edu/big-questions/how-can-astronomy-improve-life-earth

Laine, U.K. (2022) Sound-producing mechanism in the temperature inversion layer and its sensitivity to geomagnetic activity. EUROREGIO BNAM2022 Baltic-Nordic Acoustic Meeting, Aalborg, Denmark, 1-10.

Martin Nakata, Duane Hamacher, John Warren, Alex Byrne, Maurice Pagnucco, Ross Harley, Srikumar Venugopal, Kirsten Thorpe, Richard Neville & Reuben Bolt, (2014) Using Modern Technologies to Capture and Share Indigenous Astronomical Knowledge, Australian Academic & Research Libraries, 45:2, 101-110, DOI: 10.1080/00048623.2014.917786

Percy, J. (1998). Astronomy Education: An international perspective. International Astronomical Union Colloquium, 162, 2-6. doi:10.1017/S025292110011468X

Protecting and preserving traditional knowledge and culture - Indigenous Climate Hub. (2018, December 3). Indigenous Climate Hub. https://indigenousclimatehub.ca/protecting-and-preserving-traditional-knowledge-and-culture/

S Lee, A., C Maryboy, N., Buck, W., Catricheo, Y., Hamacher, D., Holbrook, J., Kimura, K., Knockwood, C., Painting, T.K &Varguez, M. (2020). Indigenous Astronomy – Best Practices and Protocols for Including Indigenous Astronomy in the Planetarium Setting. IPS Conference. https://arxiv.org/ftp/arxiv/papers/2008/2008.05266.pdf

Stewart, G. T. (2021). Defending science from what?. Educational Philosophy and Theory, 1-4.

Toledo, V. M. (1999). Indigenous Peoples and Biodiversity. Encyclopedia of Biodiversity, 3, doi: 10.1016/B978-0-12-384719-5.00299-9

Truth and Reconciliation Commission of Canada. 2015, Final Report of the Truth and Reconciliation Commission of Canada: Honouring the Truth, Reconciling for the Future. Summary. Volume One (James Lorimer Limited, Publishers)

United Nations. 2008. "United Nations Declaration on the Rights of Indigenous Peoples." Accessed March 10, 2023.http://www.un.org/esa/socdev/unpfii/documents/DRIPS_en.pdf

CONCLUSION

Readers have been taken on a journey through the rich and diverse perspectives of Indigenous peoples on astronomy and its connections to their cultures, traditions, and ways of life. From the constellations and star stories to the lunar cycles and ceremonies, and from the sun and seasonal cycles to navigation and orientation, this book has illuminated the profound and enduring relationship between Indigenous peoples and the cosmos. This book has delved into the significance of astronomy in ceremonies and explored how Indigenous astronomy intersects with modern science. In the final chapter, we emphasized the importance of preserving and sharing traditional knowledge to ensure that future generations can continue to learn from and connect with the cosmos in meaningful ways. Through the lens of Indigenous astronomy, we are reminded that the universe is not only vast and mysterious but also intimately connected to our own lives and the cultures we come from.